Decentralised Governance in Water and Sanitation in Rural India

Published by
ACADEMIC FOUNDATION
in association with

सत्यमेव जयते

NATIONAL CENTRE FOR GOOD GOVERNANCE
New Delhi • Mussoorie

Decentralised Governance in Water and Sanitation in Rural India

Editors

Kush Verma
B.S. Bisht
Aidan Cronin

ACADEMIC FOUNDATION
NEW DELHI

www.academicfoundation.com

First published in 2014
by

ACADEMIC FOUNDATION
4772-73 / 23 Bharat Ram Road, (23 Ansari Road),
Darya Ganj, New Delhi - 110 002 (India).
Phones : 23245001 / 02 / 03 / 04.
Fax : +91-11-23245005.
E-mail : books@academicfoundation.com
www.academicfoundation.com

in association with
National Centre for Good Governance, New Delhi • Mussoorie.

Cataloging in Publication Data--DK

Courtesy: D.K. Agencies (P) Ltd. <docinfo@dkagencies.com>

Decentralised governance in water and sanitation in rural India / editors, Kush Verma, B.S. Bisht, Aidan Cronin.
p. cm.
ISBN 9789332701465

1. Decentralization in government--India--Congresses. 2. Water-supply, Rural--India--Congresses. 3. Sanitation, Rural--India--Congresses. 4. Rural development--India--Congresses. I. Verma, Kush. II. Bisht, B. S., Dr. III. Cronin, Aidan.

DDC 320.80954 23

Typeset by Italics India, New Delhi.

Printed and bound by The Book Mint, New Delhi.
www.thebookmint.in

CONTENTS

List of Tables and Figures

Tables

Figures

About the Editors/Contributors

Editors:

Kush Verma, an IAS officer of 1979 batch (Uttar Pradesh Cadre), is working as Director-General, National Centre for Good Governance (NCGG), formerly known as the National Institute of Administrative Research (NIAR) under the ambit of Lal Bahadur Shastri National Academy of Administration (LBSNAA), Mussoorie. He has executed and led various assignments in Government of Uttar Pradesh as well as Government of India. He has published several papers in the field of general administration, development, labour issues and tourism etc. He has also presented papers in international/ national seminars and workshops. He is currently pursuing his PhD from Jamia Millia Islamia University, New Delhi. He has also delivered lectures at places of repute including the National Institute of Labour, NOIDA; LBSNAA, Mussoorie; Indian Institute of Public Administration (IIPA), New Delhi; Bankers Institute of Rural Development (BIRD), Lucknow etc.

B.S. Bisht is currently working as an Associate Professor at NCGG, formerly known as NIAR under the ambit of LBSNAA, Mussoorie. Prior to this, he was associated as an academician with the G.B. Pant Institute of Himalayan Environment and Development, Almora, and with the State Administrative Training Institute, Uttarakhand. Dr Bisht has a brilliant academic record and acquired his PhD from Kumaun University, Nainital. His academic interests include land-use planning, watershed management, WASH (water, sanitation and health), climate change and rural development. He has published several research papers in journals and books of national and international repute.

Aidan Cronin is trained as a civil and environmental engineer, and holds a PhD on water resources from Queens University, Belfast. He has worked in consultancy roles and then as a Senior Research Fellow at the Robens Centre, University of Surrey, UK where his research examined the impact of anthropogenic activities on water quality in the EU and developing country settings. He then worked as a water and sanitation advisor at the United Nations High Commissioner for Refugees in their Public Health Section in Geneva, Switzerland, where he reviewed the state of water and sanitation globally in refugee camps before joining UNICEF-India in 2008. He has published extensively on water, sanitation and health issues.

Contributors:

Partha Pratim Barua is working as Assistant Engineer in WSSO, PHED, Government of Assam, Guwahati.

Somnath Basu is Water and Sanitation Specialist at UNICEF, Ranchi, Jharkhand. Prior to this, he was Associate Research Director at the erstwhile Operations Research Group, currently known as AC Nielsen.

Anusha Bharadwaj is presently working as an independent Development Consultant. Prior to this, she was a Board Member at Bodhi Education Society; Manager-Resource Mobilisation & Child Rights at Gram Vikas; Social Development Specialist-School Health & ARSH at Strategic Planning and Innovation Unit, Department of Health, Medical and Family Welfare; State Coordinator at Deworm; the World Programme Coordinator, Child Rights at M.V. Foundation; Communications Coordinator at Dr Reddy's Foundation.

T.M. Vijay Bhaskar is a Joint Secretary in the Ministry of Drinking Water and Sanitation, under Government of India based at New Delhi. He is an IAS officer of 1983 batch and belongs to Karnataka cadre.

Runti Choudhury is PhD scholar at the Indian Institute of Technology (IIT), Guwahati, and a Senior Project Associate at the IIT, Guwahati.

Shashank Deshpande is a Senior Geologist at Ground Water Survey and Development Agency, Government of Maharashtra.

V.C. Goyal is the Head of Research Coordination & Management Unit in National Institute of Hydrology, Government of India, Roorkee.

Rushabh Hemani is presently working as WASH Specialist in UNICEF, Beltola, Guwahati.

Hemant Joshi is currently working as Executive Director at Communication and Capacity Development Unit, Jaipur, Rajasthan. He was Member-Secretary, Water and Sanitation Support Organisation at State Water and Sanitation Mission, and a consultant at UNICEF, India.

Yusuf Kabir is presently working as WASH Officer in UNICEF field office in Mumbai.

Jaswinder Kaur is Assistant Professor in Department of Public Administration, Punjab University, Chandigarh.

Hemant Khosla is Sanitation & Hygiene Specialist at IPE Global Private Ltd (DFID supported Swasth Project). He worked as a Consultant in WASH, UNICEF, New Delhi.

Manish Kumar is presently working as Senior Institutional Development Specialist, Water & Sanitation Program at The World Bank since January 2012. Prior to this, he was Secretary in Home and Power Department, Government of Tripura.

Namit Kumar is presently working as a Research Associate in Centre for Rural Studies, LBSNAA, Mussoorie.

Josyula Lakshmi is presently working as a Programme Manager in Centre for Good Governance (CGG), Hyderabad. She joined CGG in March 2008 with 10 years of professional experience in the development sector behind her. She is a development professional having worked with organisations such as Multiple Action Research Group (MARG), Society for Participatory Research in Asia (PRIA) and Centre for World Solidarity (CWS).

Kapil Lall (IFS) is the Project Director in Watershed Management Project, Government Region, Muni ki Reti. Prior to this, he was a Director at Project Management Unit, The Swajal Project, Dehradun.

Chicu Lokgariwar is presently working as a Content Editor in the India Water Portal (IWP), Gram Chatola, Ramgarh, Nainital. Prior to this, she was working as a Project Coordinator in Gomukh Environmental Trust (2003-2008) and then as the Head in River Conservation Group, Peoples Science Institute (2008-2011).

Chandan Mahanta is presently posted as a Professor in Department of Civil Engineering, IIT, Guwahati.

Joe Mediath is presently posted as Executive Director in Gram Vikas, Berhampur, Odisha. Gram Vikas is a non-partisan, secular, voluntary organisation working in partnership with tribal, Dalit and other poor and marginalised communities predominantly in the South and South-West Orissa.

A.K. Mishra is presently working as Consultant (KRC) at National Centre for Good Governance (erstwhile NIAR).

Vijay Mittal is presently posted as Director (Sanitation) in the Ministry of Drinking Water & Sanitation, Government of India, New Delhi.

Brecht Mommen is a Dutch national holding a Bachelors Degree in Civil Engineering and a Masters Degree in Technology and Society. He has been working as a WASH Specialist in India for UNICEF since March 2011. He has worked in the water and sanitation sector in Uganda, Indonesia, Papua New Guinea and the Netherlands.

Pravin More holds an MSc in Water and Environmental Sciences from Water Engineering Development Centre (WEDC), Loughborough University, UK. He is currently working as

WASH Specialist for UNICEF. Pravin has been practitioner, consultant and researcher with more than 25 years of experience in public policy, development, technology and planning. He has worked in multiple states including key urban centres in India.

Shantha Sheela Nair is former Secretary in the Ministry of Drinking Water & Sanitation, Government of India, New Delhi. She was the Commissioner of Chennai Municipal Corporation (1984-1988); Chairman-cum-Managing Director, Chennai Metropolitan Water Supply and Sewerage Board (1994-1997); Secretary, Government of Tamil Nadu–Urban/Rural Water Supply and Municipal Administration; and Secretary, Government of Tamil Nadu.

Ajit Phadnis is posted as Director in PriMove Infrastructure Development, Pune.

A.L. Radhakrishnan, a senior water and waste water professional, has served in Chennai Metrowater Board for the past 34 years in various capacities i.e., in-charge of planning, design, project implementation, operation & maintenance (O&M) of large urban water and waste water systems including source management, reuse, conservation and rainwater harvesting etc. Currently, he is posted as Chief Engineer (O&M) in Government of Tamil Nadu.

Suneet Sethi works as a Consultant (HRD) in Communication & Capacity Development Unit, WSSO, Jaipur, Rajasthan.

Sanjeev Sharma working as Research Associate in NCGG (erstwhile NIAR) for more than a decade has a Masters degree in Social Sciences and English and possesses a Post Graduate Diploma in Computers. He has also done Direct Trainer Skills Programme designed for trainers designed by the DoPT and conducted at LBSNAA. He has published several articles in journals and books of repute, and has been associated with successful completion of several research studies in field of education, health and rural studies.

R.D. Singh is Director of National Institute of Hydrology, Government of India, Roorkee.

Sonali Srivastava is presently working as a Consultant at Arghyam, Corporate, Bangalore. Prior to this, she was Principal Consultant at ECS Limited; Head of the Income Generation Unit at Social Work and Research Centre, Tilonia; Personnel Executive at Asian Paints India Limited.

पंकज जैन, आई.ए.एस.
PANKAJ JAIN I.A.S.

सत्यमेव जयते

सचिव
भारत सरकार
पेयजल एवं स्वच्छता मंत्रालय
Secretary
Government of India
Ministry of Drinking Water & Sanitation
247, 'A' Wing, Nirman Bhawan, New Delhi-110011
Tel.: 23061207, 23061245 Fax : 23062715
E-mail : secydws@nic.in
Website : www.ddws.nic.in

Foreword

Decentralised governance in WASH (water, sanitation and hygiene) is considered a key strategy in enhancing efficiency, equity and justice in such a vital sector that impacts on the health, nutrition and livelihood status of millions of Indians every day. Decentralisation entails the process of transferring some of the decision-making powers and responsibilities (fiscal, administrative, legal and technical) from central government down the geopolitical and administrative hierarchy to lower levels, aiming to improve the flow of information and resources between and among various levels of government.

Improving governance in the WASH sector is not an option; it is in fact essential given the enormous challenges India faces. India tragically lost more than 600,000 children under 5 years in 2010 due to diarrhoea and pneumonia,[1] almost 30 per cent of the global total. With only 4 per cent of the global water resources but 16 per cent of the world's population, India is facing serious challenges in meeting water demand. Improved governance is essential for addressing these challenges, improving efficiency and value for money, and to achieve sustainability of investments in the water

1. UNICEF (2012). *Pneumonia and Diarrhoea: Tackling the Deadliest Diseases for the World's Poorest Children*. New York: UNICEF. June. pp.86.

sector. Although the degree and nature of correlation between water sector governance and political prospects is yet to be fully explored, effective WASH governance makes good political sense to Panchayati Raj Institutions (PRIs).

A national seminar was held at the National Institute of Administrative Research (NIAR), Mussoorie, in June 2012 to explore the current status and facets of governance and WASH in India. Over 30 papers were presented by WASH experts and administrators on their experiences in the field, both in India and outside, with technical assistance from UNICEF. Existing challenges were explored and experiences shared on testing of alternate service delivery models, scaling up national water and sanitation programmes, examining community experience on acceptance, confidence and credibility in sector reform, demonstrating willingness to pay for assured, reliable and quality water services demonstrated at community level, and finally around WASH governance at local level.

This book captures the key papers from the meeting and highlights the key discussion points and potential areas for strengthening the WASH and governance processes in India. Authors are experienced sector practioners and policy makers and are well placed to contribute to this vital discourse, which this book advances.

— Pankaj Jain (IAS)

1

B.S. BISHT, C. LOKGARIWAR and
AIDAN CRONIN

Setting the Scene

Introduction

This book presents the experiences of people working towards ensuring water safety and security and sanitation and hygienic (WASH) practices for the people of India. The contributors include practitioners from government agencies, academia, international development organisations and local non-governmental organisations (NGOs). The chapters discuss the challenges in the water and sanitation sector, issues in both rural and urban areas, and the various attempts made at solving some of these problems.

The unifying thread running through these narratives—and what makes this book important—is the focus on decentralisation and governance. Decentralisation of water and sanitation programmes entails a strengthening of communities and governance mechanisms on several fronts including technological know-how, building social capital, and creating and strengthening institutions. While increasing the technical capacity of communities is essential to secure the operation of water supply and sanitation projects, it is the strengthening of governance mechanisms that has the greatest potential to create lasting change.

Access to water is crucial to life, and it is this urgency that makes the water and sanitation sector a useful entry point for development. Communities are willing to take on the responsibility of managing their own water and sanitation systems. Enabling them to do so by ensuring access to the requisite funds and functionaries creates institutions that are locally owned and managed. These institutions frequently prove to be the catalysts for further change as shown in this book.

Decentralised Governance Systems in the WASH Sector in India

Ensuring adequate access to sufficient water and sanitation to live with dignity poses a significant challenge to both governments and individuals today. The citizens of the 21st century are part of a world where, despite rapid technological breakthroughs and increased social understanding, one-sixth of the global population lacks access to safe drinking water, and nearly three times as many lack access to sanitation facilities. Poverty is a crucial factor in determining access to water and sanitation but it need not be the defining one.

What defines this access? If an individual can obtain 20 litres of water from a source less than 1 kilometer away, and a reasonably private and sanitary way of disposing of excreta and sullage, he or she belongs to the privileged two-thirds of society with access to water and sanitation [World Health Organization (WHO), 2013]. These are clearly not extravagant goals, which makes it all the more unfortunate that the world has failed to secure this basic human need. This lack of success has a debilitating effect on the lives of the poor, with increased susceptibility to illnesses including diarrhoeal diseases, schistosomiasis, trachoma, intestinal helminths and hepatitis. The first of these illnesses alone causes the death of nearly 1.5 million children under the age of five years annually.

Habitations, the smallest community unit in India, which are fully covered in terms of water access can slip back to not covered or partially covered status. Some of the official reasons for this are sources going dry or lowering of groundwater, sources become quality affected or systems working below their capacity due to poor operation and maintenance (O&M). Despite huge investments, numerous schemes and policies, the rural domestic water scenario remains quite dismal. NGO officials in the Himalayan states pointed out that though the infrastructure is in place, very few attempts have been made to ensure their sustainable maintenance (Chopra, 2010). The focus seems to be only on coverage.

This can be alleviated to a large extent by adopting a decentralised community-led approach. The World Bank's experience with Kerala Water Supply and Sanitation Project (2000-2008) and the Maharashtra Rural Water Supply and Sanitation Project (2003-2009) indicates that community-based schemes score well in terms of applicability, efficiency and quality. However, the stumbling block is in the form of long-term sustainability, especially with a lack of appropriate and strong institutions in place (Pitman, 2012).

Capacity is a key challenge to communities achieving total sanitation in India. The capacity in question is both about the techniques/technologies and the concept of community-based interventions. Both government officials and development workers find it challenging to work in the villages, around the communities' schedules. This is often offset by the rise of natural leaders within the villages, who take it upon themselves to manage the process and obtain results (Shukla, 2011).

The World Bank assisted Swajal project (1996-2003) which was implemented in 1,200 villages of Uttar Pradesh and Uttarakhand is one such example of a demand-driven approach. Villagers draw up plans based on their needs and then contribute 10 per cent towards capital cost operation and maintenance of completed schemes. The programme has largely succeeded in reducing the community's dependence on government-sponsored schemes. The appeal and success of Swajal lay in its decentralised approach to water and sanitation reforms at every level of water supply implementation. Following the success of the Swajal project, in 1999, major policy changes in rural water supply and sanitation were introduced and sector reform projects were implemented in 67 districts spread over 27 states. In 2002, a nationwide scaling up of sector reform in the form of Swajaldhara was initiated at the national level, confirming the viability of the concept. Presently, Swajaldhara along with the Total Sanitation Campaign are being implemented in all Himalayan states at the district and village level. Along with the Swajal project, there is other proof that community initiated and managed initiatives have a better chance of success than those where communities are not involved.

Strengthening Governance Systems

Following the 73rd Amendment that sought to strengthen local government bodies, Kerala embarked on an ambitious and innovative strategy for decentralising responsibility for development. The key areas of this strategy were devolving funds, functions, and functionaries into the hands of the *panchayats*, with the state government's role limited to monitoring and capacity building (Oommen, 2004). Later studies indicate that local authority approaches have been more efficient at poverty alleviation, and the creation and maintenance of infrastructure (Vijayanand, 2002).

Community-led total sanitation (CLTS) is another example of strengthening grass-roots involvement in programme implementation and is built on principles of no-subsidy and community ownership. Haryana and Himachal Pradesh, where CLTS has been implemented in all districts, have led the other states in terms of increased sanitation. During 2006-2010, the first five years of implementation, Himachal Pradesh and Haryana have demonstrated a 60-70 per cent increase in sanitation coverage. States that have partially implemented CLTS principles namely Meghalaya, Maharashtra and Madhya Pradesh have increased coverage by 30-40 per cent while Bihar with its reliance on subsidy has only increased coverage by 20 per cent (Institute of Development Studies, 2011).

This volume details several initiatives where decentralisation has contributed to strengthened governance in rural India. The impacts of decentralised governance of water and sanitation extend beyond the project goals and lifecycle. For example, increasing womens' role in water governance also improves gender relations in other areas, eventually also enabling greater financial independence (Khanna and Bisht, 2013).

Key Stakeholders, Their Concerns, Roles and Responsibilities

Domestic water supply takes precedence over all other water uses as per India's water policy. In rural areas, the 73rd Constitutional Amendment has assigned panchayati raj institutions (PRIs) the responsibility of providing drinking water; in areas that come under the Sixth Schedule, Autonomous District Councils are entrusted with this responsibility. Local governments now have a pivotal role in sanitation, O&M of water supply, and solid waste management. *Panchayats* can also contribute for rural sanitary marts, school and anganwadi sanitation, and mini water supply schemes. So far, while the legal framework is in place, certain challenges need to be overcome. These include limited accountability, inadequate representation of disadvantaged groups, insufficient clarification of roles and functions, and lack of a monitoring system. Committees have been formed at district, block and village levels. Strong committees, however, are generally only at district level.

The village water and sanitation committees have several responsibilities. These include ensuring community participation, arranging community contribution, supervision, commissioning and takeover of WASH works, collection of funds for O&M, managing and financing O&M, participation in human resource development (HRD) and information, education and communication (IEC) activities in other villages after completion of scheme in own village, and creating awareness.

In urban areas, municipal bodies and the Public Health and Engineering (PHED) departments are responsible for water supply. This largely translates as selection and tapping of sources of drinking water and creation of distribution networks based on demand. In mountain states, the department is also charged with the recurring task of repairing damage caused to the pipelines by landslides.

However, while the responsibilities have been transferred, the committees and local governance institutions are not trained to take up their roles. It is here that NGOs have a key role to play. NGOs can facilitate WASH activities at various levels including training various groups, research & development (R&D), and implementation of pilot projects. In Maharashtra, *gram panchayats* (GPs) implement projects with support from NGOs who do training and technology transfer.

Practitioners have experienced that after institution building, it is the maintenance of systems that poses a challenge. Innovative contracts including water assurance, total warranty schemes and hand pump leasing schemes may improve transparency and service provision. To do this, private sector needs to be engaged with. Some innovations recommended in this book include: adopt a life cycle approach, determine service levels, develop costs and financing overviews, strengthen accountability by decentralisation, and improve spare part supply lines.

Decentralised governance necessarily involves a diverse array of stakeholders. The three levels of rural local government, district, block and village, as well as NGOs, need to be strengthened with adequate training in management and technical skills. Within the community, it is necessary to include gender issues and the rights of marginalised groups in the planning, implementation and O&M framework. Cooperation between the government, private sector and NGOs also needs to be strengthened. Finally, water and sanitation impacts health, education and governance. It is therefore necessary to not look at WASH as a separate issue, but link it with these sectors and programmes.

WASH Sector in Rural India: Current and Future Challenges

WASH implementation, even in a decentralised approach, presents several challenges. The foremost of these are lack of awareness, distrust, little or no cooperation among the stakeholders, exclusion of marginalised communities and lack of timely release of funds. The obstacles to successful provision of water supply are difficulty of access to water supply and sanitation facilities, old projects, poor participation, lack of proper technologies, frequent damage to intakes due to rain, and need for improved assessment in provision of water services. Increased involvement of communities and a detailed assessment of availability of water resources is strongly recommended to improve the situation. Eighty per cent of rural morbidity in India is due to lack of protected and safe drinking water and sanitation. Carrying water is gender-specific with even girl children recruited to carry water during the summer months. Exclusion of marginalised groups is all too prevalent. These challenges are exacerbated by climate change; this is no longer a debatable phenomenon with its impacts being felt across the globe.

India is already coping with climate change, for example, in the Himalayan and coastal regions.

Increased involvement of communities, utilization of properly constructed infrastructure and a detailed assessment of availability of water resources is necessary for effective implementation of WASH services. Use of popular media to support interpersonal communication to promote positive behaviour change has proven useful. Convergence can be achieved by setting up coordination mechanisms, capacity building of functionaries and identifying of key components for convergence with strong monitoring for impact. While suitable technology needs to be applied, it is now being realised that it forms only part of the solution. In order for communities to make informed choices, institutional transformation and awareness raising is necessary as well as handholding of the communities to achieve their aims.

The Way Ahead

Ensuring access to water and sanitation is an essential first step towards achievement of other goals for peoples' well-being, such as health and livelihood. It is increasingly evident, as per the chapters that follow, that decentralised governance offers a huge potential for achieving this goal equitably and sustainably. Decentralised systems that are demanded, initiated, planned, implemented, maintained and financed by the communities involved are not only efficient and sustainable, but also meet ancillary goals such as gender parity.

At the same time, there is a gap between communities that lack technical knowledge and institutions that lack the social skills to competently set up sustainable decentralised water and sanitation systems. An element of disconnect also may exist between these two entities. This book presents case studies where an attempt has been made to bridge this gap with increased transparency and institution building, notably Srivastava (Chapter 14). A concentrated effort to implement a similar strategy throughout the country is called for. This capacity building needs to take place on several fronts: technical aspects of designing, constructing and maintaining a water supply/sewerage systems; the management aspects of creating and sustaining institutions; the science skills required to test water quality and adjust treatment methods accordingly; and social skills such as conflict resolution. Establishing policies and programmes that strive to bridge this knowledge gap is essential to make efficient decentralised governance a reality in the country today. These policies must then be implemented with the most marginalised of the community in focus with a service delivery timeline in mind and with quality standards in place, an even bigger challenge.

References

Chopra, R. (2010). *Water Resources in the Himalayan Region*. Report prepared for the Planning Commission, Government of India. Peoples Science Institute.

Institute of Development Studies (2011). *Community led Total Sanitation: India*. Online URL: *http://www.communityledtotalsanitation.org/country/india?page=1*

Khanna, P. and B.S. Bisht (2013). *Participatory Water Governance through Gender Sensitisation in the Indian Context*. Submitted for the 3rd Sustainable Mountain Development Summit. Kohima. September 2013.

Oommen, M.A. (2004). "Deepening Decentralised Governance in Rural India: Lessons from the People's Plan Initiative of Kerala", *Working Paper No. 11*. Centre for Socio-economic & Environmental Studies.

Pitman, George T. (2012). *Project Performance Assessment Report (Republic of India) Kerala Rural Water Supply and Environmental Sanitation Project "JALANIDHI" (CREDIT 3431-In) and Maharashtra Rural Water Supply and Sanitation Project "JALSWARAJYA" (CREDIT 3821-In)*. World Bank.

Shukla, J.P. (2011). *Major Challenges in doing CLTS in India*. Online URL: *http://www.communityledtotalsanitation.org/blog/major-challenges-doing-clts-india*

Vijayanand, S.M. (2002). "Poverty Reduction through Decentralisation: Lessons from the Experience of Kerala State", Poverty Net Work Home Page.

World Health Organisation (2013). *Health through Safe drinking Water and Sanitation*. Online URL: *http://www.who.int/water_sanitation_health/mdg1/en/*

2

T.M. VIJAY BHASKAR

Decentralised Governance in Drinking Water Sector

Introduction

The Ministry of Drinking Water and Sanitation administers the National Rural Drinking Water Programme (NRDWP) and the Total Sanitation Campaign, through which support is extended to the states for implementing rural drinking water supply and sanitation schemes.

The achievement of the programme is measured by the number of habitations, where the population is fully covered with adequate (40 lpcd) and safe drinking water. As per information received from the states, through the online monitoring system of the ministry, the coverage of habitations is about 72 per cent of total rural habitations. The rest are either partially covered or have some chemical contamination problems in their drinking water sources. The achievement as estimated by independent studies like the National Sample Survey Organisation (NSSO), 65th Round Survey of 2008-09, shows that about 90 per cent of rural households obtain their drinking water from improved sources. The coverage through piped water supply, however, is still poor at about 35 per cent. But the slippage of habitations from fully covered to partially covered due to depleting groundwater sources, contamination of sources, poor operation and maintenance (O&M) of schemes, poor community or *panchayats'* involvement leading to poor service delivery continues to be problematic. Accordingly, the sector requires further addressing of the issues related to the local governance to effectively provide the services to rural households.

Defining Decentralised Governance

Decentralisation is the process of dispersing the decision making governance, closer to the people. The authority of decision making is to be shifted down to the lower levels of the system, to facilitate the flow of decision making from bottom to top and so also of the ideas. Some basic controls must be in place for this process to be successful. Such decentralisation will be successful and effective only when the capacity of the decision making is also correspondingly developed and the decision makers have access to adequate information and also have basic professional guidance to enable these institutions to take proper decisions.

The basic concept of decentralised governance is to build the capacity of the social system to mobilise available energies in a coherent way, in this case, for the sustainable development and O&M of the drinking water supply systems and in the larger context of all water resources. In order to be effective, sustainable and acceptable to the local population, such governance must be transparent, open, participatory, communicative, coherent, efficient and above all, accountable. There would be problems in the beginning during the transformation, depending on the customs, understanding of the issues and the previous social order.

Decentralisation of governance to the *panchayats* or at the village level, based on the principle of subsidiarity, is critical to the success of these programmes, as the problems can best be identified at these lowest levels only. If adequate authority is provided to the institutions existing at these levels, or created as per specific needs of the sector, the identified problems can be addressed and can be solved, with the required technical support to these institutions.

The transfer of responsibility has to be associated with the transfer of authority for carrying out the sector activities or public functions, such as the management of the water resources or the O&M of the water supply schemes. When we devolve the functions along with authority and responsibility, we must also transfer the finance and the functionaries to the decentralised governing body or institution to facilitate its smooth functioning.

Many local factors will also decide the way forward for a successful model of decentralised governance. So far as the water sector is concerned, many a times, non-governmental organisation (NGOs) act as catalysts to help the community, to participate in various issues and thus take advantage of the change. The political systems also affect the success of the decentralisation process. There is increased probability that the good performance of the functionaries of the decentralised governing institutions, such as *panchayats*, in service delivery, can help and determine the re-election of these representatives.

Progress and Challenges

India tragically loses more than 600 children under five years each day on account of dehydration due to diarrhea (Black and Liu, n.d.), 88 per cent of these deaths are linked to inadequate water and sanitation service provision. Furthermore, an additional 980 children die in India every day due to pneumonia which is also strongly associated with poor personal hygiene (Fewtrell *et al.*, 2007). India though has made very good progress on providing access to improved drinking water sources for its population and the number of people who gained access to improved drinking water sources from 1990 to 2010 was greatest in India with 522 million [World Health Organization (WHO)/United Nations International Children's Education Fund (UNICEF), 2012]. Key drawbacks with these figures are that it does not reflect quality of water (only access to an improved source) and UNICEF-supported research work shows that water quality issues in India remain a very serious challenge, as per Government of India (GoI) communication to states.[1] In terms of access, there are still 97 million people in India accessing drinking water from unimproved sources.[2] India is not on track to meet the sanitation Millennium Development Goal (MDG) and tops the global list of open defecators (626 million people in total), though India is continuing to accelerate sanitation users with

1. GoI letter to States on February 2, 2011, reference W-11042/01/2011/Water; Subject Report on Multi-District Water Assessment by UNICEF.
2. Ibid.

an average 21 million new toilet users per year during the period 2000-2010.[3]

Another key issue behind these statistics is that of who is gaining access: 65 per cent of the richest quintile of India have piped water on premises while it is only 2 per cent of the poorest quintile; in rural areas 32 per cent of the richest quintile have piped water on premises while it is 1 per cent of the poorest quintile. Also interestingly, 17 per cent of scheduled tribe (ST) households have access to water on their residential premises whereas it is 35 per cent for scheduled castes (SC), 43 per cent for other backward castes (OBCs) and 49 per cent for all other categories.[4]

Central Level Initiatives

The evolution of the Rural Drinking Water Supply Programme in the country has seen many phases since independence. In 1950, the Constitution of India specified 'Water' as a State subject. In 1969, the NRDWP was launched. It was changed to the Accelerated Rural Water Supply Programme in 1972-73 to assist the states to accelerate the pace of supplying safe drinking water to the rural population. The National Drinking Water Mission was launched in 1986 to further accelerate the process of coverage of rural habitations with safe water supply; it was renamed as Rajiv Gandhi National Drinking Water Mission in the year 1991. The first National Policy was drafted by the Ministry of Water Resources in the year 1987 which gave top priority to drinking water supply. In 1994, the 73rd Constitution Amendment was passed for assigning the responsibility of providing drinking water to the panchayati raj institutions (PRIs). In 1999, a separate Department of Drinking Water Supply was created in the Ministry of Rural Development and a Sector Reform Programme was started, for a paradigm shift from the supply-driven approach to a demand-driven approach. Total Sanitation Campaign was also initiated to ensure proper rural sanitation with the goal of eradicating the practice of open defecation.

---

3. Ibid.
4. National Family Health Survey-3, 2005/06, GoI.

The Water Supply Programme was further scaled up in 2002 with addition of the demand-driven Swajaldhara Programme in which the PRIs were given funds through the States and made to assess their demand, plan, design, implement operate and maintain their own water supply schemes, based on their actual needs, keeping in mind the sustainability of such schemes.

The country also committed itself to the MDGs to half the proportion of rural population without sustainable access to safe drinking water and basic sanitation by 2015, from the 1990 levels. The ambitious Bharat Nirman Programme was also launched in 2005 to provide drinking water to all uncovered, slipped back and quality affected habitations. A Sub-Mission programme for providing safe drinking water to quality affected villages was also started this year. In 2009-10, the Accelerated Rural Water Supply Programme (ARWSP) was modified as the NRDWP, mainstreaming the Swajaldhara principles of demand-driven implementation of schemes in a participatory manner. The Department of Drinking Water & Sanitation became an independent Ministry in 2011.

State Level Initiatives

Until recently, there were two basic models for implementing the rural water supply programme at the state level, the Public Health Engineering Departments (PHED) under the direct control of the state government or an Autonomous Water & Sanitation Board. Although these institutions have done a commendable job in terms of coverage of the habitations with safe water supply, they suffer from some technical and/or management problems such as poor hydro-geological investigations; a relatively high percentage of improper designs, which led to substantial cost overruns; and limited quality assurance, despite a comprehensive set of built-in controls that appear to be inherent in the underlying supply-driven 'design and construct' character of these sector agencies, responsible for rural water supply delivery.

The 73rd Amendment to the Constitution seeks to endow the rural *panchayats* with such powers and authority as may be required to enable them to function as effective and genuine institutions of self-governance. As Panchayati Raj is included in the state list of the 7th Schedule of the Constitution, the states are responsible for devolution of powers to the *panchayats* and thus need to take steps for effective devolution of powers and authority as envisaged in Article 243G, read with the 11th Schedule of the Constitution. Accordingly, the GoI strongly advocates and also incentivises states for the implementation and maintenance of the Rural Drinking Water and Sanitation Programmes by the PRIs, with technical support of the line departments of the states.

Investment in the Sector

From the 1st Plan period to the 11th Plan (2007-2012), a total investment of about ₹ 1.56 lakh crores has been made in this sector by the centre and the states together. Despite this huge investment, every rural person in the country is yet to be provided safe and adequate drinking water at all times of the year and there are several challenges being faced by this sector, the main being lack of effective governance and taking over the responsibility by the village institutions and the communities.

Issues and Challenges

About 28 per cent of rural habitations are yet to be fully provided with safe and adequate drinking water supply. As per the Census 2011, about 22 per cent of rural households have to travel more than 500 meters to fetch drinking water. However, as per NSSO 65th Round Survey 2008-09, only 3 per cent got it beyond 500 meters. Only 30.8 per cent of rural households have piped water supply and an even lower 12 per cent have household tap connections (NSSO 2008-09). About 35 per cent of rural households get water within the household premises either from taps, wells or hand pumps. Service levels of water supply in terms of days of supply in a year or supply hours in a day are yet to be measured. The NSSO 65th Round Survey reported that 13.8 per cent of rural households did not get sufficient drinking water during one or more months of the year. The most difficult month was May when nearly 12.5 per cent rural households did not get sufficient drinking water with June and April being the other most difficult months.

More than 85 per cent of the rural water supplies schemes are based on groundwater in the country, as these are comparatively cheaper and elaborate water treatment systems are not required, only disinfection is generally adequate. Groundwater availability is not uniform, there are regional imbalances in the availability of groundwater. There is high demand of the agriculture sector on groundwater and its over exploitation is very adversely affecting the sustainability of water supply schemes.

In the case of the rural water supply and sanitation sectors, the *panchayats* have had the responsibility for sector operations and maintenance thrust upon them, while the PHEDs have not retreated from the position of asset owners and purveyor of technology and service delivery. Thus while the 73rd Amendment provides the needed policy and legislative framework for devolving responsibility for rural water supply and sanitation to the PRIs, the devolution of management and financial management responsibility to the *panchayats* has a long way to go.

The current institutional arrangements of water resource management do not enable comprehensive water allocation. There is also lack of coordination between various institutions. At the same time, there is duplication of responsibilities and gaps in accountability. There is inadequate fostering of the grass-roots level institutions. There has also been lack of private sector participation or involvement; this is also true for the involvement of the PRIs and the rural communities or user groups. The rural drinking water sector has also been suffering from inappropriate pricing policies, water resource management policies in general and its strategies. Reforms are needed in institutional arrangements and mechanisms; strong legislative and regulatory frame works are needed; economic and financial incentives and mechanisms are also required.

Rural drinking water supply schemes have suffered mainly from poor upkeep. Although the responsibility of O&M of the schemes lies with the PRIs, it is poorly defined in many states and not supported by transfer of adequate funds and trained functionaries to the PRIs. These PRIs and the Village Water and Sanitation Committees, formed as per the National Policy for rural drinking water, are generally not willing to take over completed water supply schemes in which they were not involved during the planning and implementation process by the concerned departments. Inadequate water resource investigation, improper designs, poor construction, sub-standard material and workmanship, and lack of preventive maintenance of these schemes lead to rapid deterioration of these rural water supply schemes.

These complex issues related with water resources management can be addressed in an effective way only at the local level, through decentralised governance.

Need and Status of the Regulatory Framework

Traditionally, groundwater has been considered as part of the land, so the owner of the surface property has had an absolute and exclusive right to use it. Gradually, as the interdependence of surface and groundwater is confirmed, together with the possibility of groundwater depletion, protective measures are being included in the legislation.

Modern legal frameworks tend to regulate the use of groundwater resource either through the public domain or by exercising the administrative powers. State regulation is a necessary consequence of the modern exploitation technologies related to water. The earlier technologies did not involve large-scale assaults on the water resource. With the limited exploitation, the impact was marginal and as a result, legal regulations were not essential. Modern technologies allow large-scale exploitation which also affects the water quality adversely. In these circumstances, the need of suitable legislation is acceptable. However even in states where groundwater development laws have been enacted, their effective enforcement remains an issue. In this context, self-regulation by the community and *gram panchayat* (GP), backed by legislative teeth, may show the way out.

Andhra Pradesh Farmer Managed Groundwater Systems Project

A very successful example of decentralised governance for self-regulation of groundwater abstraction is the Andhra Pradesh Farmer

Managed Groundwater Systems (APFAMGS) project, based on the basic principle of user's understanding of groundwater occurrence, cycle and availability. Sustainable management of groundwater is done through the formation of the Groundwater Management Committees (GMC) at the Hydrological Units (HU) level. The GMCs are trained to monitor groundwater and considering its availability and water budgeting, the community makes changes in the cropping pattern.

Groundwater has been a major source of irrigation, meeting more than half of the irrigation needs of Andhra Pradesh. There are more than 2.2 million wells including open wells, infiltration wells and tube wells. Large parts of the state are covered by hard rocks where aquifers are not extensive and offer limited discharge. The depth of the tube wells is steadily increasing from 30 mts to 90 mts and even up to 200 mts in some areas. Over exploitation of the groundwater is leading to steep decline in groundwater levels and failure of many tube wells in many parts of the state, but the farmers are still investing in new tube wells.

Under these circumstances, the APFAMGS has demonstrated an alternative model that encourages active involvement of local communities in basic data collection, capacity building and improved management of the available groundwater resources.

The overall objective of this project is to demonstrate an alternative strategy for managing groundwater distress using the demand side management approach, focusing on strong people's institutions and enhancing the knowledge base of the community. A key strategy of the project is that community members are encouraged to collect data on groundwater and cropping systems, build on their existing knowledge and develop reflective capacity to consider the aspects of improved crop water use. The strategy is to use knowledge (both local and in public domain) as the main input to get a proper perspective for managing the groundwater decline. APFAMGS experience shows that when knowledge levels are enhanced, enlightened communities act unitedly to make difficult choices, willingly.

APFAMGS project, initiated in 2004, has successfully demonstrated that through decentralised governance and involvement of community through improved knowledge, behavioural change can be brought about. Enlightened communities adopt voluntary self-regulation, leading to reduced groundwater pumping and improved water use efficiency. In seven drought-prone districts of Andhra Pradesh, thousands of farmers residing in hundred of habitations, spread over several hundred kilometres, are voluntarily taking number of steps to reduce groundwater pumping, for tiding over the problem of groundwater depletion.

Pilot Drinking Water Security Project

A similar effort in the rural drinking water and sanitation sector has recently been initiated by the Ministry of Drinking Water and Sanitation by launching a Pilot Drinking Water Security Project, in collaboration with 10 states, in 15 over exploited blocks. The basic objective of this project is to achieve drinking water security through convergence and participatory integrated water management. The three pronged approach adopted for the implementation of this project includes firstly the implementation of source sustainability measures through convergence with other programmes such as Mahatma Gandhi National Rural Employment Guarantee Act (MGNREGA), Integrated Water Resources Management Programme, Participatory Water Shed Development Programme and the Ministry's own NRDWP through active net working. The second approach is the participatory integrated water resource management by strengthening the capacity of local governments and community institutions including the GP for groundwater monitoring and self-regulation of demand. The third approach is the preparation of drinking water security plans by the GPs.

The solution to the problems of water quality and competing demands of water use from other sectors, mainly agriculture, cannot be found in isolation by the drinking water sector alone. It needs to be addressed in a holistic manner by involving the users and empowering them to plan for the water resource management to meet their various needs while ensuring drinking water security.

This pilot project is aimed at creating awareness among the stakeholders, not only in understanding the challenges of water security but also in finding the solutions through proper planning and water resource management and also demand management. The learnings from this pilot project will go a long way in achieving drinking water security and adequate sanitation facilities in rural India.

Incentivising States for Devolution

As a key strategy to achieve the goal of providing adequate safe water to all the rural households, strengthening of decentralisation of governance is being promoted by the Ministry of Drinking Water and Sanitation. The states have to ensure that the rural water supply schemes are planned, implemented, operated and maintained by the community, and for this, institutionalisation has to be done through the formation of Village Water and Sanitation Committees. As a very powerful tool, incentive fund is being provided to the states to encourage and incentivise the process of decentralisation and for this, a Management Devolution Index (MDI) has been designed to realistically calculate the extent to which this devolution has actually been achieved by the states, based on which, the incentive fund is allocated to the states.

In the inter-state allocation criteria for allocating NRDWP funds, there is a weightage of 10 per cent for 'rural population managing rural drinking water supply schemes' to encourage state governments to devolve management of rural water supply schemes to PRIs. The earlier method for allocating this incentive fund was based on the figures of rural population managing rural drinking water supply schemes, obtained from all states. The funds were allocated to each state in proportion to the rural population managing water supply schemes in that state, as a proportion of the total for all states.

This method did not measure the depth or extent of management devolved to the PRIs. At present, while some states have transferred the full range of functions to GPs, others have only transferred a few functions. Some states have transferred only hand pumps management to GPs, some have transferred single village piped water supply schemes and some have even transferred multi-village water supply schemes. Flow of funds to the *panchayats* remains a problem in many states and functionaries support to the *panchayats* is also a weak area. The net result is that due to lack of substantive devolution, the *panchayats* are handicapped in managing the schemes leading to poor O&M of the schemes, non-functionality and poor service delivery.

Management Devolution Index

It was therefore proposed to refine this method to promote more substantive devolution of functions, funds and functionaries to the *panchayats*. To this end, the MDI has been designed, based on selected indicators that measure the depth of management devolution to PRIs for more effective allocation of the fund.

The MDI covers all the related aspects of transfer of functions, funds and functionaries to the PRIs.

Under the transfer of functions, the indicators include the status of issue of State Acts or orders which clearly define the transfer of responsibility for infrastructure creation of hand pumps, single village and multi-village water supply schemes; actual proportion of these schemes where the O&M has been transferred to PRIs; formation of the Village Water and Sanitation Committees as sub-committees of the GP under State Act or rule; the status of testing of the drinking water sources, etc.

Under the transfer of funds, the indicators include proportion of central and state funds (both for creation and for O&M) transferred to the accounts of PRIs or *zila panchayats*; transfer of plan and non-plan state grants to GPs; transfer of Finance Commission grants to GPs; electricity tariff for water supply schemes as compared to the lowest slab for domestic users in the state and also the percentage of user charges collected by the PRIs against their demand.

To ensure adequate support of the functionaries, the indicators included to calculate the MDI are: the status of engineers made available at all the blocks exclusively for drinking water and sanitation activities; proportion of the posts of consultants in the DWSM and those of BRCs at block level filled up; VWSC members trained; NRDWP support activities funds spent on the information, education and communication (IEC) and human resource development (HRD) activities. Good management indicators of proportion of the households given private service connections and the status of metering in the villages where bulk water supply has been given are also included.

Each indicator is given a weight between 2 to 15. The highest weight-age of 50 per cent is given for indicators measuring transfer of funds. Marking is done for all these indicators based on the progress achieved in the state as per documents submitted, and based on the total MDI score earned by the state and the population managing rural water supply schemes in the state, and the incentive fund is allocated.

A tabular statement showing the MDI scores, the population managing rural water supply schemes and the amount allocated under 10 per cent incentive fund state-wise is given in the Annexure.

Conclusion

Decentralised governance in the rural drinking water sector will effectively address many of the issues plaguing the sector. It will help in the protection of the environmental use, by defining levels of eco systems protection and water pollution control, while protecting the interests of disadvantaged sections and customary users. Conflict resolution will be easy, particularly for the conflicts relating to the competing demands of water.

In the initial phase of decentralised governance, the governments should stimulate and facilitate the participation of interested villagers, in providing access to information, and provide a forum for expressing opinions. It is also necessary to ensure that the privileged user groups do not have a dominant position.

The decentralisation also allows the possibility of experimentation, in the way a given public service is provided and a better way of governance is demonstrated. Such efforts will also lead to technological and institutional innovations at the village level. Such decentralised, local governance will also have better accountability. By transferring of the control from the top to the bottom, service delivery will also improve, as the authority will go to those who will be closer to and more responsive to user needs. It would be possible to have a better pricing structure for rural drinking water services with appropriately designed cross-subsidy structure. However, in the rural context, cross-subsidisation may not be the only solution as the number of users benefiting from the subsidies may be higher than those financing the system and therefore, government's financial support will be needed to make the systems sustainable.

It is generally recognised that the drinking water supply and sanitation services are characterised by significant economies of scale. Decentralised governance will take this aspect also into account and will be able to join with neighbouring *panchayats* to increase the service areas, if that is technically possible.

Drinking water security can be best ensured by building capacity of local communities to monitor and measure their water resources, prepare water budgets and take steps to self-regulate demand for water for irrigation and industry. Therefore, aquifer management by villages and *panchayats* should be an integral part of ensuring drinking water security especially in water stressed and quality affected areas.

During the course of decentralisation, the ground realities that must also be considered when designing community involvement strategies for rural water supply and sanitation (RWSS) is that the social organisation in Indian villages is often very heterogeneous with different caste groups and large disparities between rich and poor. The basic social constraints such as high levels of illiteracy in some states especially among women and the poor should also be kept in mind. Furthermore, the size of local governments, including GPs, varies significantly in the country. As the smaller sized *panchayats* are easier to plan for and easier to administrate, they are less viable; as larger units require more administrative and organisational infrastructure, those are more viable.

All such issues will be effectively dealt with at the local government level through community participation. This participation of the beneficiaries in water supply schemes should be ensured right from the planning stage, spanning over construction and post-scheme completion management stages, including O&M. Decentralisation of the governance at the *panchayat* or at the village level will be a key to the success of rural drinking water and sanitation programmes in the country.

References

Black, Robert E. and Li Liu (n.d.). *Global Child Mortality: Estimates of Levels and Causes for 2000-2010* (preliminary results). Bloomberg School of Public Health, Johns Hopkins University.

Fewtrell, L., A. Prüss-Üstün, R. Bos, F. Gore and J. Bartram (2007). *Water, Sanitation and Hygiene: Quantifying the Health Impact at National and Local Levels in Countries with Incomplete Water Supply and Sanitation Coverage.* World Health Organization Environmental Burden of Disease Series No. 15. pp.71.

WHO/UNICEF (2012). *Joint Monitoring Programme for Water Supply and Sanitation.*

Annexure Table A-2.1

State-wise MDI Scores, the Population Managing Rural Water Supply Schemes and the Amount Allocated under 10 per cent Incentive Fund State-wise

S.No.	States	2012-13 (MDI Allocation)					
		Marks Obtained by States	Rural Population as per 2011	Rural Population as per 2011	Population Managing	Management Devolution Index	Funds to States based on MDI (in lakh)
1	2	3	4	5	6	7	8
1	Andhra Pradesh	30.45	5.6E+07	56311.8	0	0.00	0.00
2	Bihar	7.78	9.2E+07	92075	72635:2	565101.86	3488.16
3	Chhattisgarh	23.03	2E+07	19603.7	16795	386788.85	2387.50
4	Goa	0	551414	551.414	0	0.00	0.00
5	Gujarat	30.36	3.5E+07	34670.8	14298	434087.28	2679.46
6	Haryana	13.9	1.7E+07	16531.5	3380.87	46994.09	290.08
7	Himachal Pradesh	15.82	6167805	6167.81	1031	16310.42	100.68
8	J&K	3.75	9134820	9134.82	0	0.00	0.00
9	Jharkhand	22.85	2.5E+07	25036.9	1665	38045.25	234.84
10	Karnataka	66.25	3.8E+07	37552.5	38350	2540687.50	15682.72
11	Kerala	22.92	1.7E+07	17455.5	17455.5	400080.20	2469.55
12	M.P.	14.47	5.3E+07	52537.9	9982	144439.54	891.57
13	Maharashtra	32.62	6.2E+07	61545.4	56765.4	1851688.49	11429.78
14	Orissa	16.05	3.5E+07	34951.2	5385.1	86430.86	533.51
15	Punjab	23.48	1.7E+07	17316.8	7982	187417.36	1156.86
16	Rajasthan	40.36	5.2E+07	51540.2	3158	127456.88	786.74
17	Tamil Nadu	48.44	3.7E+07	37189.2	33607	1627923.08	10048.56
18	Uttar Pradesh	19.02	1.6E+08	155111	155111	2950211.64	18210.56
19	Uttaranchal	27.75	7025583	7025.58	779.8	21639.45	133.57
20	West Bengal	16.97	6.2E+07	62213.7	58481	992422.57	6125.86
	Total		7.9E+08			12417725.31	76650.00

contd...

.contd...

1	2	3	4	5	6	7	8
DDP							
1	Andhra Pradesh	30.45			0	0.00	0.00
2	Gujarat	30.36			14298	434087.28	1439.85
3	Haryana	13.9			3380.87	46994.09	155.88
4	Himachal Pradesh	15.82			1031.23	16314.06	54.11
5	J&K	3.75			0	0.00	0.00
6	Karnataka	66.25		.	38350	2540687.50	8427.38
7	Rajasthan	40.36			3158	127456.88	.422.77
	Total					3165539.81	10500.00
North-East							
1	Arunachal Pradesh	12.8			70.105	897.34	181.41
2	Assam	13.79			529.117	7296.52	1475.12
3	Manipur	11.31			274.223	3101.46	627.02
4	Meghalaya	5.19			57.915	300.58	60.77
5	Mizoram	10.84			305.2	3308.37	668.85
6	Nagaland	17.52			1406.86	24648.20	4983.08
7	Sikkim	20.24			455.962	9228.67	1865.74
8	Tripura	21.52			146.625	3155.37	637.91
	Total					51936.52	10500.00
	Grand Total					15635201.64	97650.00

3

AIDAN CRONIN and HEMANT KHOSLA

Perspectives on Governance and Sustainability

Introduction

The recent GDP growth is a real reason of pride for India but behind this there are two questions: (i) how it can be sustained within the current challenges of water safety and security? and (ii) how to spread the benefits to all segments of the population? Indeed, Goldman Sachs (2008) places improving environmental sustainability of water and energy as one of the top 10 interventions that India needs to accelerate in order to reach its economic potential. Water is, therefore, without doubt, central to sustained and equitable growth of India. However, only with equitable policy that is implemented well in a solid governance environment can these twin goals be achieved.

Evolution of the Rural Drinking Water Supply Sector

The Government of India (GoI) has always placed immense importance on rural water supply and it is worthwhile to quickly trace the evolution of the sector in India (Figure 3.1). Four phases of programmes can be traced out.

Phase 1: The First Five Year Plan (1951-1956) allocated approximately 1.5 per cent of the total budget to water and sanitation. The allocation remained roughly the same until 1969 when the National Rural Drinking Water Supply Programme (NRDWP) was launched with technical support from the United Nations International Children Education Fund (UNICEF) with emphasis on drilling bore wells. The Fourth Five Year Plan (1969-

1974) saw a near doubling of budgetary allocation. During this plan, the Accelerated Rural Water Supply Programme (ARWSP) was launched in 1972. This programme aimed to increase the pace of drinking water supply coverage which had a sluggish pace until this time. Budgetary allocation for water and sanitation also peaked at 4.15 per cent under the Sixth Five Year Plan.

Phase 2: A national level apex committee was set up in 1981 to define policies that were envisaged to guide India to provide access to safe water in all its villages. This led to formation of the National Drinking Water Mission (NDWM) in 1986 and was followed by the first National Water Policy in 1987 by the Ministry of Water Resources. Additionally, a Technology Mission was launched in 1986-87 in which appropriate technology intervention, water quality, human resource development and support were introduced in rural water supply sector. The Technology Mission was later renamed as Rajiv Gandhi National Drinking Water Mission (RGNDWM) in 1991.

Phase 3: 2002 was a very important year in the India water sector. The National Water Policy was revised in 2002 and villages that did not have adequate sources of safe water were prioritised and the level of service delivery in partially covered villages was targeted for improvement. Also in 2002 India committed to Millennium Development Goals (MDGs). However, from the governance and operation and maintenance (O&M) perspective, the year 2002 also saw the launch of the Swajaldhara programme which changed the way in which water and sanitation services were supported in rural areas. Swajaldhara advocated decentralised service delivery by devolution of powers to the local government systems, i.e. PRIs (panchyati raj institutions) and users. Under this new approach, communities have been bestowed with the powers of planning, implementation, ownership and O&M of water supply schemes. For the first time, users agreed to pay an upfront tariff for water use, that was set significantly higher to cover the O&M cost. Swajaldhara continued until 2009 when the NRDWP was introduced. Under Swajaldhara, the community was supposed to take up 100 per cent of O&M in addition to sharing 10 per cent partial capital cost in cash or kind. However, line department officials were slow to adopt reforms and the PRIs lacked finances and skills to take up the responsibility immediately.

Phase 4: The rural water supply sector today appears remarkably different in its approach and prioritisation of issues, in comparison to the earlier generations. A significant difference lies in the shift from top-down, technology-driven approach to a decentralised one which lays greater emphasis on aspects like sustainability, equity and community participation. Ensuring sustainability of water and its availability in terms of portability, adequacy, convenience, affordability and equity are given higher priority in the latest programmes. All this is to be achieved while ensuring decentralised approach which involves PRIs and community organisations. A major push towards decentralisation of powers and responsibility was given with the 73rd and 74th Constitutional Amendment which assigned the responsibility of drinking water to PRIs. Department of Drinking Water Supply was renamed as the Department of Drinking Water and Sanitation in 2010, which later was upgraded to the Ministry of Drinking Water and Sanitation (MDWS) in 2011. MDWS has developed a 10-year strategy[1] for water in which it advocates the key linkages between sustainability, quality, O&M and good governance. The challenge still remains to implement such forward-looking policies and strategies across a country the size of India with 600,000 villages.

Indeed, the evolution clearly shows the emphasis moving away from only capital cost outlays to an increasing focus on sustainability and O&M, and this has been a positive development from financial and sustainability points of view, but then this means the human capacity needs (at local and higher levels) must be deliberated on. Equally important is the process to achieve this—both in terms of empowering communities, especially women and minorities, and *via* uptake of the equally important software to lead to sustained behavioural change.

1. Ministry of Drinking Water and Sanitation Strategy 2012-2022, MDWS *website www.ddws.nic.in.*

Figure 3.1

Rural Water Supply and Sanitation in India over the Years

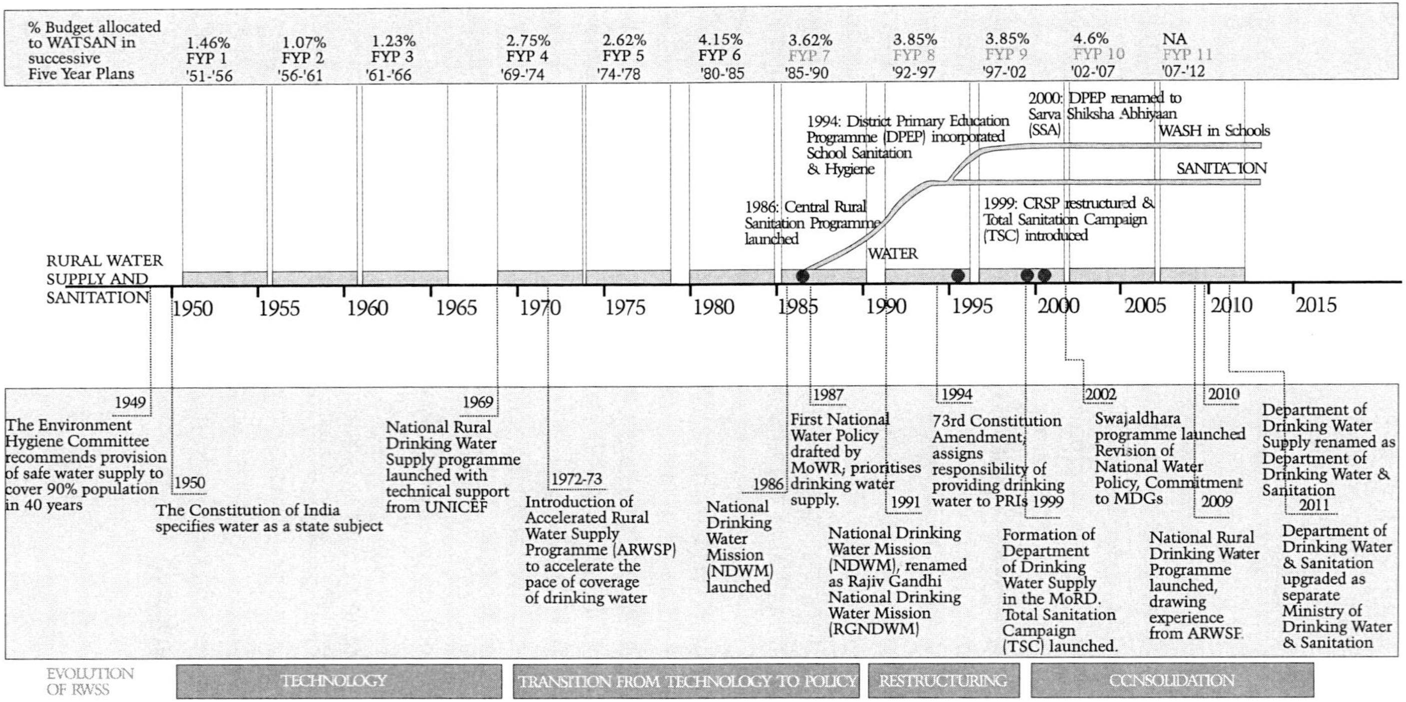

Note: A broad trend analysis of the timeline indicates that restructuring of various programmes and departments happened in the period 1989-1999 which was followed by consolidation of these different institutions and their programmes, post 2000 (UNICEF, 2012).

The Current State of Rural Water Infrastructure Operation and Maintenance

NRDWP has provision of 10 per cent of the annual outlay for O&M of water supply schemes. Under the Sector Reforms and Swajaldhara programmes, the GoI advocated the 10 per cent capital cost sharing by the PRIs/communities and 100 per cent O&M being borne by the PRIs/community through collection of tariff from the user households. Furthermore, to aid the financial costs and providing financial assistance to the PRIs successfully managing their water supply schemes, there are provisions of incentives up to 10 per cent of the cost of schemes to be transferred to the villages. Suitable policy directions and guidelines have been specified and laid out by the MDWS in the NRDWP to aid the successful O&M being carried out by the user group/community. Over and above the financial assistance towards successful O&M, the MDWS has also initiated the process of incentivising the states for better O&M. A Management Devolution Index (MDI) matrix (Annexure Table A-3.1) developed by MDWS for this purpose captures the efforts of state government for devolution of powers to the PRIs/community. There have been successful instances of water supply schemes in states such as Gujarat and Tamil Nadu and some other states with notable effects.

In the case of a single village water supply, a major concern for expanding water-supply and sanitation services is to select technologies and institutional options that users would be willing to pay for, and that would also ensure good public health and sustainable environmental conditions. Further, the O&M systems that are developed by the community not only depend on the water supply system, but also on the involvement and support of the community. Several issues need to be addressed before users are willing to pay O&M charges (WASMO, 2006a). In the community-managed projects facilitated by Water and Sanitation Management Organisation (WASMO) the following conditions are key criteria to be met:

- Demand generation for the facilities.
- Generating the need for O&M.
- Creating demand for better service delivery.

- Technical capacity of the user community to carry out O&M.
- Ability of the community to devise appropriate water tariff.

O&M of systems involve recurring costs and costs for regular and emergency repairs. The maintenance and repair formula prescribed by the Government of Gujarat takes into account structure maintenance, establishment, chemicals/materials, payment to supplier (in case of transfer of water by the government) and energy charges (presently subsidised by the government). The tariff is designed in such a way that the maintenance and repair cost is recovered from the users, but the cost for replacement is not expected from the community.

Multi-village schemes for drinking water supply are again more complicated. At the district-level, the water supply schemes are designed and implemented through a system of engaging contractors for the implementation of the schemes. Post-implementation, these schemes are either being fully managed by the respective Public Health Engineering Department (PHED) or through a process of out-sourcing of the schemes for regular O&M by contractors which are selected through a tendering process.

A variety of factors affect the O&M of the multi-village schemes which include:

- Capacity and ability of the contractor to carry out regular O&M.
- Adequate trained manpower and capacity building of human resource available.
- Technical support by the PHED to the contractors for regular O&M.
- Willingness of the community to repay the user charges.

Defining Good Governance and Combating Corruption in WASH

Governance is the act of governing. Good governance has eight major characteristics. It is participatory, consensus-oriented, accountable, transparent, responsive, effective and efficient, equitable and inclusive and follows the rule of law. It assures that corruption is minimised, the views of minorities are taken into account and that the voices of the most vulnerable in society are heard in decision making. It is also responsive to the present and future needs of society.[2] Lack of governance is usually manifested by practices ranging from procedural irregularities to illegal and criminal activities. Indeed, it has been estimated that globally corruption adds $50 billion onto the price tag of achieving the MDG on water and sanitation (Transparency International, 2008) Three main types of corruption can be found in the rural water supply sector: procurement irregularities (e.g., over-invoicing on officials tenders), leakages from development budgets (i.e., payments to senior bureaucrats and politicians) and unviable distortions of planned schemes (reworking of schemes due to political or other pressures that may be against technical best practices and will eventually compromise the scheme functionality) (James, 2010).

At the core of these issues are some fundamental points that need to be addressed. Firstly, corruption flourishes when the focus in not on results but on 'leakage of funds.' Hence corruption thrives when the outcomes of schemes (both short-term and long-term) are not measured and so independent monitoring is key to combatting corruption. The results of such monitoring have to be linked to accountability of officials involved in such schemes. Power brings legal responsibilities for service delivery and such action against such responsibilities should bring reward or censure as appropriate. Three types of audit are essential for such monitoring. Financial auditing is in practice, but when it is done alone, it cannot comment on value for money or on quality of service delivery. It must be linked with a social audit to reflect on what has been delivered to the community. However, even these combined are not enough—a third prong is required—technical audit. A technical audit is required in parallel with financial audits to show value for money, sustainability and safety of the scheme in the long term. The authors argue that it is the lack of this three prong approach of combining financial, social and independent technical auditing that has resulted in many well-intentioned

2. The United Nations Economic and Social Commission for Asia and the Pacific (ESCAP), accessed April 2012.

initiatives at curbing corruption not realising their full potential (e.g., Right to Information Act, 2005, community agitation, public interest litigations, etc.) in terms of water safety and security and sanitations and hygienic practices (WASH) sustainability.

It is not enough that the drinking water systems work on the first month or first year of commissioning. Rural water schemes must deliver safe water equitably and sustainably.[3] Good O&M is a manifestation of good governance—when this happens, the likelihood of sustainability and safety of water is much higher. Hence, planning from the outset must address a life cycle approach that costs schemes under a combination of capital costs, installation costs, preventative maintenance costs, reactive maintenance costs, running costs over a period of say 10 years. An example is given from Bangladesh and surprisingly shows that the 10-year per capita complete costing varies from $7 to $35. This underlines the need for extensive community consultation as with a factor of difference of only 5, any solution that is not owned and embraced by the community is a wrong choice. India may also think of the creation of such a spreadsheet showing typical ranges of these costs in India, eventually disaggregating to the per cubic metre costs.

Building an Effective and Participatory WASH

A study[4] examining the reasons behind why rural people in India do not demand safe water interestingly found that:

- People do not clearly distinguish between water that appears clean and water that is safe. If water tastes sweet, looked clean and transparent and was free from bad odour, it is considered safe.
- The main concern of people continues to be availability of adequate quantity of water rather than availability of safe water. In cases where water is available, the focus of the villagers is to ensure that the source is near and accessible.

3. Secretary MDWS at International Workshop on "Mitigating Effects of Geogenic Contaminants (MEGC)" held at CSIR-NEERI, Nagpur, February 22, 2012.
4. Sample Study on Why Rural People in India do not demand Safe Drinking Water, Ernst &Young for UNICEF, December 2010.

Table 3.1

Comparison of 10-Year Life Cycle Costing or Rural Water Supply Options in Bangladesh

Technology	*# of People Served*	*Capital Costs ($)*	*Annual O&M Costs ($)*	*Cost/10 yrs @ 10% d.r. ($)*	*10 yrs Capital Cost ($)*
Deep Tube Well (DTW)	50	744.6	10	806.0	16.12
DTW Tara-Dev (DTW-T)	50	893.1	14	979.1	19.58
Rain Water Harvesting (RWH)	50	855.3	25.5	1,012.0	20.24
PSF (PSF)	50	558.8	80	1,050.4	21.01
Dug Well (DW)	50	772.6	80	1,264.2	25.28
Pond Sand Filter (30 HH)	150	558.8	80	1,050.4	7.00
DTW-Multiple Connection (DTW-MC)	200	1,344	40	1,589.6	7.95
Mini Piped Deep Well (GOB)	500	9,000	1,350	17,295.2	34.59
Piped Deep Well (DPHE)	2,500	43,285	6,780	84,945.2	33.98
Piped River Abstraction	2,500	44,533	6,780	86,193.2	34.48
Piped Impoundment	2,500	48,244	7,627	95,108.6	38.04

Source: Presented by Dr Yan Zheng, UNICEF Dhaka, at the India Water Quality Exposition, November 15, 2011.

- There is very limited awareness about safe water as a right.
- Even where there is demand for water, a lack of positive response from the service providers usually leads to acceptance of the existing situation. Lack of redressed mechanisms is another major hindrance to improved service.
- Personal interaction with service providers can be a more effective way of asserting reliability of a message than ones delivered through wall paintings, radio, mobiles or TV.

This shows that an effective and sustainable WASH sector requires a practical demand responsive approach and this requires solid governance initiatives to ensure that such a process is planned and followed through. States such as Gujarat and Tamil Nadu may have developed such an approach *via* WASMO and Tamil Nadu Water Supply and Drainage (TWAD) Board, but perhaps for water supply only and have omitted sanitation to some extent.

The crucial element of sanitation has been seen as less important, perhaps because it is more difficult, takes a longer time and requires fewer resources. Key characteristics of an effective strategy could include (Cronin and Burgers, 2011):

- The community initiates and makes informed choices about service options, based on their willingness to pay for the service, and acceptance of responsibilities for operation and maintenance.
- The community contributes to investment costs relative to the level of service and has control over how funds are managed.
- There is an adequate flow of information to the community, as well as procedures for facilitating collective action decisions within the community
- Communities can choose how goods and services are delivered and how water and sanitation services are managed.
- District and state governments create an enabling environment, while the federal government sets clear national policies and strategies.
- The community owns and is responsible for sustaining its facilities.
- Community capacity is appropriately strengthened.
- Innovation is promoted and the need for flexibility is recognised, especially for equitable supply to all segments of the population.

Community engagement at all stages is crucial and they are the key stakeholders to ensure quality, impact, sustainability and equity in rural water supply. A detailed understanding of this process and its relationship with hardware, human resource requirements and costing is essential to optimise interventions. A Water Quality Monitoring Framework developed by Argyham[5]

5. A framework document for good water quality management, collating experiences from the voluntary sector, Argyham, 2012, pp.88.

and its partners builds up such a structure. It is based on detailed feedback from NGOs across India to map out what is required at all stages of a water intervention to ensure that it succeeds. The structure identifies five key stages in the process, from baseline assessment, to participatory planning, to water quality monitoring and data management to implementation and finally O&M. A first interesting reflection is that implementation is actually at step 4; it is also interesting that hardware costs are in total only about half of the project costs. Also interesting is the importance of building up social capital (and in all other senses) in villages as part of the process. Their experience has shown the entire project cycle takes 2.5 to 5 years and any short-cutting of the processes. Such a structure is, in fact, cause and effect of good governance of the rural water sector in India. Other successful examples in India, such as WASMO's community managed drinking water supply programmes in Gujarat or Gram Vikas in Odisha, show that up to two years are required to establish basic level of services and followed up with continuous and institutionalised support from higher levels thereafter. This has been shown to eventually culminate in sustainable infrastructure, use of toilets and proper hygienic practices. Lala *et al.* (forthcoming) have documented approaches on WASH and gender in India and again have found the needs for a multi-year approach to achieve such goals. Hence, process and structure are extremely important in creating demand and sustainability and can be key reflections of good governance in systems.

Promoting an Enabling Operation and Management Environment

Increasingly, Governments, external support agencies and local communities are recognising the importance of integrating O&M components in all development phases of water-supply and sanitation projects, including the planning, implementation, management and monitoring phases (Brikké and Bredero, 2003). The GoI plays a vital role in creating an 'enabling environment' within which an O&M policy framework can be developed, one of the key elements of sustainability. Such an environment can be fostered in a number of ways, including through legal provisions, regulations, education initiatives, capacity building initiatives and by communicating information. An important role of local government is to promote an awareness of national policies and to support community water-user committees. Both the implementing agencies and the community should be made aware of the O&M implications, as the communities themselves have responsibilities in the management and O&M of their water-supply and sanitation systems. However, as many PRIs/communities have insufficient resources and/or capacity, support by the local government may also be influenced. The roles and responsibilities of the actors involved in O&M need to be well defined, especially where governments are shifting from their traditional role as a service provider to that of a facilitator of service provision.

Sector professionals use a variety of terms to describe affordable, simple technologies that can be adapted to local conditions. Water supply and sanitation projects should not be viewed as an end in themselves, but as the initiators of benefits that continue long after the projects have been handed over to the community. However, to ensure that long-term benefits are realised, the projects must be sustainable, which means appropriate technologies must be selected, and O&M should be integrated into project development from the beginning. Although community-based projects may take longer to develop than short-term, externally managed projects, the longer development time can be used to identify factors that would influence service sustainability. Often, critical aspects of O&M development have been neglected in short-term, externally managed projects. Effective O&M brings about important health benefits by sustaining accessible water supplies in adequate quantity and quality; by reducing the time and effort spent on water collection; by allowing better sanitation facilities to be provided; and by providing income-generating activities.

Communities are often prone to skepticism about developmental projects, often when it comes to 'participation'. Several steps need to be taken to increase transparency in project implementation, which would culminate into effective O&M by user groups (WASMO, 2006b). Transparency measures can include, for instance, decision making in village meetings, *gram sabha*, proper maintenance of records, display boards for assets created, constitution of vigilant committees and regular social auditing of project works. Public access to information as enacted by the government under Right to Information Act may also lead to transparent transactions and dealings. Selection of contractors, vendors, hired/outsourced agencies through process of e-tendering; display of information through government websites; development of grievance redressal systems; and citizen's charter may lead to more effective and transparent monitoring of projects.

The role of the private sector must also be discussed. There has been a tendency to decentralise O&M activities and to encourage the private sector to get involved in both the construction and upkeep of water-supply and sanitation facilities. Although this trend could increase the flexibility of O&M activities and reduce costs, private sector involvement may be limited by the current low profit margins, particularly in areas where rural communities are scattered and poor. Private-sector accountability is also a concern when there are little controls or regulations. Government agencies and/or communities that contract services from the private sector need to ensure that the job is well done at a fair price. To some extent, the communities themselves can monitor the quality of the work, even though they may initially require assistance from the government agencies. Nevertheless, informal community-based monitoring is no substitute for developing government guidelines to ensure there are minimum-quality standards for the work and such interventions are cost-effective. It is also important that the guidelines be conveyed to the communities, since they have increasing responsibilities, not only in the O&M of their water-supply systems, but also in their financial management. Regulation, control and monitoring require extensive efforts and commitment by governments, and considerable human and financial resources.

Ultimately it is how all of this translates down to the marginalised and unreached that will determine the success or failure of governance in the WASH sector in India. Disaggregated data are essential to assess this, an example is the district-based WASH data validation that was completed in 12 districts of Gujarat to examine the disparities in the status of WASH facilities at household and institutional levels; it also examined the usage of facilities by men, women and children (Mudgerikar and Cronin, 2012).

Conclusions

Tremendous progress has been made in the Indian water sector, including the achievement of the MDG target for drinking water. Despite this, serious health-related burden still exists in India linked to poor water and sanitation and almost 100 million still lack access to improved drinking water source in the country. The sector has seen significant policy and financial outlay progress in the rural water sector since independence. The Swajaldhara programme and the NRDWP have been advancing policy to look at community participation, contribution and ownership of schemes and their role in rural water asset O&M. This issue of community engagement is seen as one of the most critical elements of O&M and good governance in the sector leading to equitable, safe and sustainable water supply for all. Unfortunately, this is still not fully achieved in India and corruption is hindering progress in this respect also. This can be thwarted by transparent processes in implantation, balancing social, technical and financial audits but above all with strong independent monitoring and a clear focus on results as opposed to outlays or process indicators. Structured approaches to all phases of planning, implementing, monitoring and reviewing can help in this respect. Monitoring of the impact of interventions with gender and equity viewpoints can perhaps be the best indicator for success in good governance in the rural WASH sector in India.

References

Brikké, François and Maarten Bredero (2003). *Linking Technology Choice with Operation and Maintenance in the Context of Community Water Supply and Sanitation*. Geneva, Switzerland: World Health Organization and IRC Water and Sanitation Centre.

Cronin, A.A. and L. Burgers (2011). "Water Safety and Security—Challenges and Opportunities in the Indian Perspective, in Confluence of Ideas and Organizations", *International Conference on Water Partnerships towards Meeting the Climate Challenge*. Chennai. January. pp.45-52.

Goldman Sachs (2008). "Ten Things for India to Achieve its 2050 Potential", *Global Economics Paper* No. 169.

James, A.J. (2010). "Challenging the 'Percentage System', Improved Transparency and Accountability in Rural Water Supply Programmes in India", in *International Symposium on 'Pumps, Pipes and Promises'*, November 16-18. The Hague, Netherlands: IRC International Water and Sanitation Centre.

Lala, S., M. Basu, Jyotsna and A.A. Cronin (forthcoming) "Frameworks for Accelerating Improved Gender Outcomes for the WASH Sector in India", Chapter 2, in *Gender & WASH in India*. SAGE Publications.

Mudgerikar, A. and A.A. Cronin (2012). "Review of the Status of Equity in WASH Programming in India", Paper prepared for *The South Asia Sanitation and Hygiene Workshop*. January 31 to February 2. Rajendrapur, Bangladesh; edited by Kathleen Shordt, Carmen da Silva Wells and IngeborgKrukkert.

Transparency International (2008). *Global Corruption Report*. *www.transparency.org*.

UNICEF (2012). *WASH in India Handbook*, 62.

WASMO (2006a). *Users as Managers—Facilitating Local Self Governance or Drinking Water Supply*. Water Supply Department, Government of Gujarat.

———. (2006b). *Initiatives for Social audit, Good Governance & Transparency*. Water Supply Department, Government of Gujarat.

Disclaimer: The views expressed herein are those of the authors and do not necessarily reflect the views of UNICEF or the United Nations.

Annexure Table A-3.1

Management Devolution Index, MDWS

S.I. No.	*Devolution Indicators*	*Weightage (%)*
1	Transfer of Functions to PRIs	
1.1	The state acts and/or executive orders/MoUs should clearly define the transfer of responsibility for infrastructure creation of hand pumps to PRIs.	2
1.2	The state acts and/or executive orders/MoUs should clearly define the transfer of responsibility for infrastructure creation of single village piped water schemes to PRIs.	2
1.3	The state acts and/or executive orders should clearly define the transfer of responsibility for O&M of hand pumps to PRIs.	2
1.4	The state acts and/or executive orders should clearly define the transfer of responsibility for O&M of single village piped water supply schemes to PRIs.	2
1.5	Proportion of hand pumps whose O&M has been transferred to PRIs.	2
1.6	Proportion of single village piped water schemes whose O&M has been transferred to PRIs.	2
1.7	The VWSCs should be standing/sub-committees of GPs under the state act/rules.	2
1.8	Proportion of drinking water sources for which water quality testing done during the previous years as per the IMIS.	6
	Total for Functions	20
2	Availability/transfer of funds to PRIs.	
2.1	Proportion of NRDWP (coverage and quality) funds (central + state share), transferred to PRI/DWSM subordinate to ZP accounts.	15
2.2	Proportion of NRDWP (O&M) (central +state share) transferred to GPs accounts.	15
2.3	Whether untied grants from state plan/non-plan have been transferred to all GPs in preceding financial year.	5
2.4	Whether untied grants from state finance commission have been transferred to all GPs in preceding financial year.	5
2.5	Whether unit charges of electricity for pumping in drinking water supply schemes by PRIs equal to or lesser than lowest slab of unit charge for domestic consumers.	5
2.6	Percentage of water charges demand collected by PRIs.	5
	Total for funds transfer to PRIs	50

contd...

...contd...

3	Functionaries made available to support PRIs.	
3.1	Proportion of blocks where block level diploma/graduate engineers are available in PHED/PRED/ZP/BP @ one for 1 lakh rural population exclusively for rural water supply and sanitation.	5
3.2	Proportion of filled up DWSM consultants positions.	5
3.3	Proportion of filled up BRC coordinators positions.	5
3.4	Proportion of VWSC members trained in RWS functions for at least two days.	5
3.5	Proportion of NRDWP support funds spent on IEC and HRD activities.	5
	Total for functionaries' support to PRIs	25
4	Good management indicators.	
4.1	Proportion of households in state provided with household connections.	2
4.2	Proportion of villages with bulk water supplies metered.	3
	Total for good management indicators	5
	Grand total	100

Note: GP: *gram panchayat*; VWSC: village water and sanitation committee; IMIS: Integrated Management Information System; DWSM: District Water and Sanitation Mission.

4

SHANTHA SHEELA NAIR and
A.L. RADHAKRISHNAN

Utilising Water and Improving Sanitation in Chennai and Musiri

This chapter considers two cases of urban water management. The first is Chennai, the capital of the Indian state of Tamil Nadu, one of the country's four big metropolises. It has a population of over 5.6 million and has been infamous for its acute scarcity of drinking water and huge unsustainable systems, which relied on tankers and pipelines to transport drinking water long distances. The city exploited the groundwater under it and its neighbour's groundwater too. The other case is Musiri Town Panchayat, which had no town sanitation system, just a plethora of informal, inadequate and unhealthy ways of dispensing with human waste. Besides lacking funds, the town faced environmental difficulties in integrating conventional sanitation systems.

The need for *in situ* solutions at affordable costs, and more importantly sustainable solutions, led Chennai to harvest rainwater from rooftops and recharge rainwater into groundwater aquifers. This proved more lasting, wholesome, sustainable and environment friendly. Thus this chapter explores a potable water-supply model for large urban cities to consider. It shows the significance of active citizens' participation in a wholly green-centric paradigm moving from water scarcity to water security. The case of Musiri Town Panchayat involved convincing householders to accept using waterless toilets and teaching the community how to use them appropriately.

Many waste management systems—especially waterborne sewerage systems, which were hailed as the public-health miracle of the industrial revolution—are no longer sustainable either ecologically or economically in burgeoning cities, especially of the developing world. With greater knowledge of ecological principles and the understanding that urban waste should not follow 'put out of sight' or 'flush and forget' systems, human waste management has assumed a strong significance in urban planning. This chapter discusses two Indian cases of meeting basic human waste need in the face of distinct ecological challenges, Chennai and Musiri. Note that the Chennai City and Metrowater (Chennai) websites, referred to at the end, are the sources for all the statistical and other material.

Chennai: From Water Scarcity to Security

All over the world metropolitan, water supply system managers have valued using more water per capita from increasingly distant sources, leading to crises. Ironically, the inefficient design of such systems without regard to urban resources and constraints has aggravated drought-flood cycles. Harnessing rainwater and floodwater using urban space as a catchment is a striking example of environmentally sustainable service provision. A unitary drought-proofing flood-mitigation plan is a dream, given the current urban nightmares; the first tale, of Chennai, is one of water scarcity to water security.

Chennai, a small fishing hamlet on the Bay of Bengal in the 17th century, was developed by the British East India Company into a trading post known as Madras. Recently restored its original name, Chennai is the fourth largest metropolis and capital of Tamil Nadu, the southernmost state of India. It spreads over 76 km^2 of extremely flat coastal plains. The Kosathalaiyar, Cooum and Adyar rivers criss-cross the city and empty into the Bay of Bengal. Buckingham Canal, a large waterway, runs north-south. The population was estimated to be over 5.6 million in 2009. Chennai has an average annual rainfall of around 1250 mm, with more than 75 per cent falling during October-December.

Until about 1870, the people of Madras had depended on shallow household wells and on public wells and neighbourhood tanks. An organised system began in 1872 and formed the nucleus of Chennai's civic surface water supply system to this day. The civic authorities followed the principle of developed world countries by maintaining water supply at per capita levels. More than a century later, with several augmentation measures combining both surface and groundwater, in the year 2000, the supply was around 350 megalitres per day for a population of nearly five million.

However, in a bad monsoon year, even 200 megalitres per day proved difficult to supply. Averaged out, drinking-water sources consisted half of surface water and half of groundwater. However, in normal monsoon years, 80 per cent came from surface-water reservoirs and 20 per cent from groundwater reserves. In bad monsoon years, all drinking water came from groundwater, as lakes went dry and surface water could not be used.

The increasing drinking-water crisis in the growing city led to the formulation of master plans to access water from rivers, which required interstate negotiations and agreements, and obtaining water through canals or pipes running up to 200-300 km. A rising demand for water and ever-decreasing per capita availability led to excessive dependence on groundwater by private enterprises, individuals, households and the civic authorities.

As mining and selling water increased alarmingly, groundwater aquifers became saline and fresh-water aquifers became depleted. Overdependence on groundwater and its excessive withdrawal led to the intrusion of sea water in the north-eastern coastal belt: an ingress only 3 km inshore in 1969 increased to 7 km in 1983 and up to 9 km in 1987. With diminishing water quality, private water vendors started 'mining' water from the southern coastal aquifer. Groundwater levels of the city fell and brackish water appeared in some localities which had good quality groundwater earlier. The seriousness of the problem was obvious when Chennai witnessed acute droughts in the years 1983, 1987, 1993 and again from 2001 through to mid-2004. Recurring droughts, almost every alternate year, led to thoughts about the draconian measure of evacuating urban areas due to water scarcity.

The gravity of the water crisis prompted the Chennai Metropolitan Water Supply and Sewerage Board to revise its master plan to manage its water supply systems more sustainably. Besides augmenting drinking water by identifying distant surface sources, unique features of the plan involved regulating groundwater and conserving water through decentralised, *in situ*, affordable and sustainable solutions. The state and city legislated to prevent uncontrolled commercial mining and over-extraction of groundwater. The Government of Tamil Nadu introduced the Chennai Metropolitan Area Ground Water Regulation Act in 1987, amending it in 2002.

Figure 4.1

Changes in Ground Water Levels for Chennai (1983-2007)

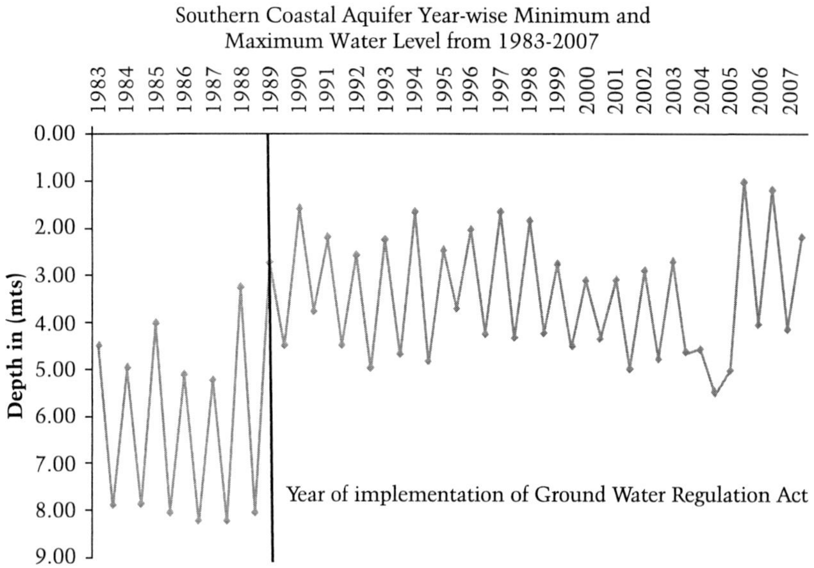

Source: Senior Hydro-geologist, Chennai Water.

The city started wastewater recycling for individual and non-potable uses. The legislative and water-diversion measures contributed to improvements in normal years. However, in the event of droughts, both individual citizens and the civic body still depended on groundwater. As droughts were widespread, even water supplies from distant sources became unreliable, had to be shared and, in crises, were invariably retained by upper riparian populations. Persistent conflict forced authorities to find better, internal, solutions.

The answer was tapping the potential of the almost 1,250 mm rainfall which the city received in normal years but which was received during a short duration of four months, invariably leading to inundation and floods, disrupting regular water supplies and sewerage systems. Due to its flat topography, floods took days to recede. Drainage was impeded further throughout the monsoons by rising tides and high sea levels. It was felt that tapping this rainfall could augment drinking-water supplies and could also be used to charge the groundwater aquifer.

Small efforts were made to make rooftop rainwater harvesting mandatory in building rules in the metropolitan development authority's town planning requirements. In 1994, rainwater harvesting systems were mandated for permission to build multi-storey and special buildings. In 2000, such systems became mandatory in all buildings applying for new water and sewerage connections, but they received a very lukewarm response and were limited to large, newly constructed buildings. In 2003, systems to harvest rainwater were made compulsory in all buildings. The state legislature amended municipal and local laws giving owners one year from the date of notification to provide a system in every building.

The rainwater harvesting systems so promoted had three aims: to collect water from urban roofs; to filter water so harvested for consumption through pre-existing sumps; and to encourage appropriate filtering of excess water and run-off (e.g., from streets) for charging into the ground and the prolific urban wells—both open and deep-bore.

With the decentralised availability of drinking water, the paradigm shifted from a system wholly controlled by engineers and the Chennai Metropolitan Water Supply and Sewerage Board to a level of co-management. Households accessing simple water supply systems was an innovative intervention in drinking-water management, while authorities maintained large professionally controlled and formal water supply systems.

The success of these measures depended on unrelated but concerted actions for implementing the process 'in letter, spirit and functionality of the law.' A major step involved citizens' groups and print and electronic media in an intense awareness programme to demystify the harvesting of rainwater and argue for its water quality and sustainability. Technical data were provided to residents, associations, civic groups and decentralised civic units. Rainwater harvesting models and rain collection centres were developed in various localities.

Further awareness and information were promoted through government and civic bodies undertaking the harvesting of rainwater from their buildings. Politically important persons, including the governor and chief minister of the state, went public on the benefits they received from adopting rainwater harvesting systems, which had an instant appeal to the common person. Celebrities, including cricketers, musicians and film stars, were actively involved in the campaign too. The campaign was strengthened by a discreet but effective message that non-adoption was punishable.

Rainwater harvesting enhanced the availability of safe and hygienic drinking water and became an important, novel urban feature. Besides meeting the gap in demand and supply, the city saves ₹ 300-400 million annually. Households save on water charges to the water agency during monsoons and to private suppliers at other times.

The low-lying flood-prone areas became a source of water instead of the usual scourge after a downpour. Floodwaters were guided into aquifers through simple filters. The entire programme had drought-proofing and flood-mitigation features. There has been a perceptible decrease in sea-water ingress. Environmentalists and green groups ensure that these 2003 reforms are retained and functional for every monsoon.

The design paradigm adopted by Chennai—decentralised water supply systems for more conservative, more personalised water supply management systems to suit the household and the community—has led not only to less wastage of water but also in yet to be calculated savings in energy otherwise needed for water and sewage treatment. Decentralised water collection and usage is more democratic as well as more ecologically sustainable than conventional centralised water supply systems.

Rainwater harvesting gave Chennai an alternative to establishing desalination plants. Providing such services with people's participation is must be considered by all large urban cities while planning and managing their drinking water. Chennai serves as a precedent and model for other similarly challenged cities. The experiences of Chennai and the measures taken are typical of responses to a crisis. A city that was infamous for its acute scarcity of drinking water and exploitation of groundwater is moving towards water security achieved by application of affordable and environmentally sustainable solutions created *in situ* and with social benefits.

Intensive Waste Management in Musiri

Unlike Chennai, the small municipality of Musri Town Panchayat, in Tamil Nadu, had plentiful water but no sewerage system because of problems selecting an appropriate one. However, revolution in a small town was more feasible. It involved accepting waterless toilets. Thus a decentralised urban waste disposal system was introduced and proved less costly and more environment friendly than conventional models.

Traditionally, urban sanitation models depended on large volumes of water. A typical city sewerage system pumped solid and liquid human waste into massive slurries, which were transported miles into rural hinterlands for treatment and disposal. This was costly and risked ecological disaster as untreated and partially treated sewage polluted water bodies and rivers. River cleaning proved economically unviable and ecologically untenable.

The Musiri population of 35,000 spreads over 15.6 km^2. Of 6,200 odd households, around 3,500 have septic tank latrines, 1,200 leach pit latrines and the rest use community latrines or the surrounding countryside. Situated on the banks of the river Cauvery, the town has a unique topography of criss-crossing natural drains, irrigation channels and a high water-table due to its proximity to the river. Open rains fed into various irrigation channels, groundwater was

Figure 4.2

Rainwater Harvesting Techniques

Rainwater Harvesting in Thatched Roof

Rainwater Harvesting in Sloped Roof

Rainwater Harvesting in Sump and Well

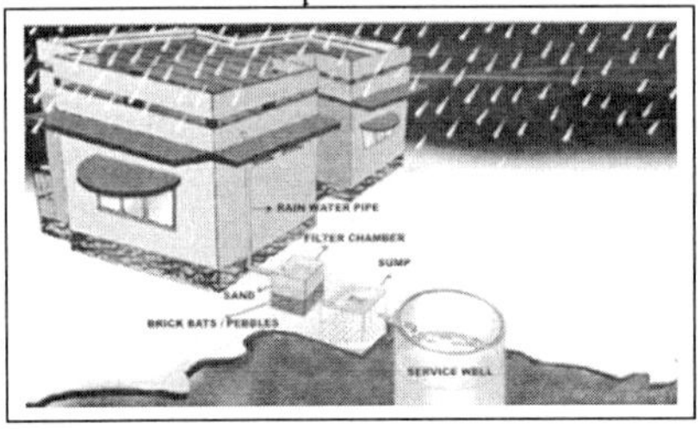

Rainwater Harvesting in Multi-storied Building

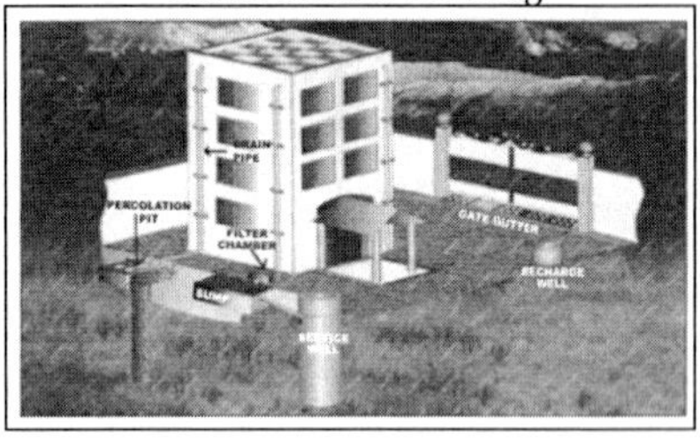

Road Side Rain Water Collection through Recharge Well

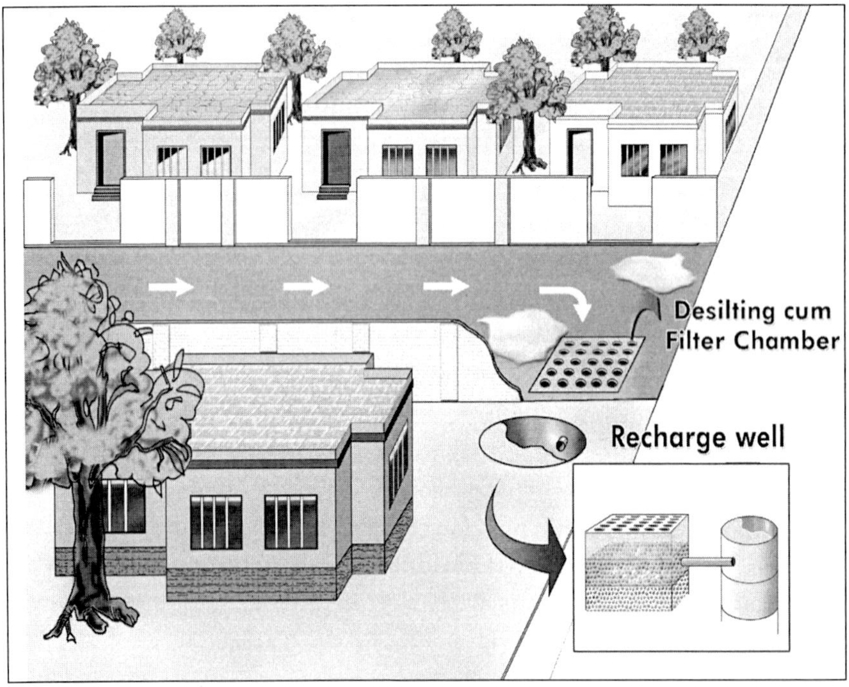

close to the surface and faecal matter leached from pit toilets, overflowing and polluting groundwater and well water.

Musri had enough water but could not afford a conventional sewerage system, so it willingly entered a partnership to implement an ecologically safe sanitation system. *In situ* treatment of waste had found favour in neighbouring rural areas when the elected town body, supported by a local non-government organisation Society for Conservation & Protection of Environment (SCOPE), and the state government, decided to pilot models of urine diversion and *in situ* drying of human faecal matter.

To translate this vision into action, the Musiri Town Panchayat (supported by the state government) together with SCOPE conducted a baseline survey to assess the current sanitation infrastructure facilities and people's level of awareness related to health and hygiene. The survey showed that most households had leach pit toilets and septic tanks, which polluted the groundwater and also polluted the Cauvery river. As most land was agriculturally fertile, there was no wasteland available for large sewage disposal systems. The existing community toilets were dilapidated and many people defecated in the open. There was also fear that any 'acceptable' system that used water would contaminate the town's sources, including drinking water, hence there was a general acceptance to the concept of a waterless system.

The first challenge was to convince the townsfolk to adopt a waterless system for faecal disposal while using water for after-defecation washing. The second challenge was to show the potential for re-use of urine and faeces in an ecologically safe and economically wise manner. The third challenge was to upscale from household to community models by designing and setting up public latrines and urinals with their many inherent hygiene and maintenance issues.

The consortium decided to construct household and community toilets based on the ecological sanitation (EcoSan) model. To address the problems of disposal of solid and liquid waste, an integrated decentralised wastewater treatment system and solid and liquid waste management plans were also undertaken. 'EcoSan' is an approach that contains, sanitises and recovers human excreta for use in soil systems to enhance agricultural production. EcoSan comprises soil-based composting toilets in shallow reinforced pits for the faecal matter and urine diversion with or without storage vaults. Some designs also mix the wash water with urine. 'Closing the loop' on sanitation and water is shortened and waste reused through monetary incentives. EcoSan models protect water resources by reducing consumption and contamination. Recovery of nutrients for use to improve agricultural yields also minimises water-based infections.

The first household-centred EcoSan toilet model designed in Musiri was based on the original developed in Thanneerpandal village training centre in 2002. In the two-in-one model, urine and washing water were collected in a mud pot outside the toilet. Faeces were collected in a chamber for composting. The successful demonstration model motivated Mrs Mangalathammal of Kaliyapalayam village to construct the first EcoSan private household toilet and, in turn, some 170 families from the nearby villages of Kaliyapalayam and Seventhilingapuram constructed similar household toilets.

Based on feedback from the local community, the EcoSan was redesigned into a three-in-one model in tune with the prevalent Indian practice of a drop hole in the middle, washbowl in the rear and urine bowl in the front. The washing water, urine and faeces do not mix and are collected separately. Nutrients recovered from urine and washing water collected in the toilet are used for irrigating vegetables and flowers in household kitchen gardens and community farms.

As the EcoSan model and its usage and maintenance practices were novel and unfamiliar, an intensive awareness programme was organised for users and elected representatives. The purpose of the different pipes, twin chambers and features of the squatting slab were laboriously explained to ensure ease and comfort of use. Do's and don'ts were painted on the toilet's inner walls in the local language. All members of the community, especially women's self-help groups (SHGs), were actively involved in planning, designing, construction, operation and maintenance of the EcoSan toilets.

Figure 4.3

The EcoSan Model

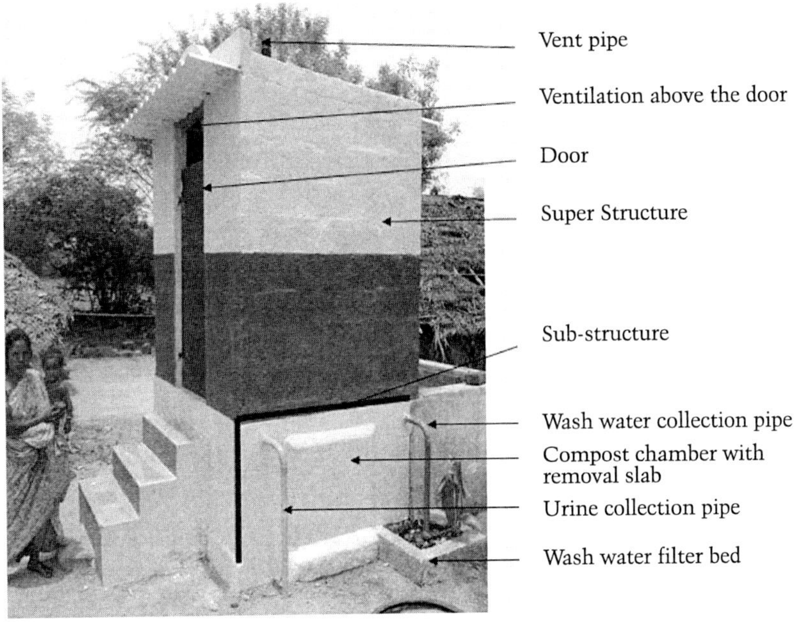

Intensive training for masons was undertaken. A training module and manual on EcoSan community compost toilets (ECCTs) was developed and their operation and maintenance were learned by SHG members and town sanitary workers. To ensure sustainability in the post-construction phase, SCOPE ensured continuous monitoring with the community. Families were approached to assess problems of use, availability of biomass/ash/sand to cover the faecal matter, the correct uncontaminated diversion of urine for best results and the filtering of washing water through simple systems before use for growing flowers or vegetables. Solutions were meticulously followed up. SCOPE even organised a beauty contest for EcoSan toilets to promote their use, maintenance and awareness.

Figure 4.4

Construction of an EcoSan Toilet

The baseline survey had found that two community toilets were dilapidated, forcing a large number of people to defecate in the open and on the riverbed. It was decided to construct an ECCT in Saliyar Street which was perhaps India's first ECCT.

Again there was meticulous designing of the size and number of chambers, pipes conveying water and urine to the filter bed and urine collection tank, the pans and notices on use practices, along with the concept of EcoSan. The project had tremendous support from the highest district officials, with the foundation being laid by the District Collector of Tiruchi in the presence of potential users and residents of Saliyar Street.

The ECCT was built with bricks, with a 90-square-foot compost chamber. The redesigned three-in-one pan was used, faeces being collected and dehydrated in the chamber and urine collected in a tank outside. The washing water directly drained into a filter bed with banana trees. The attractive toilet had walls and floors of glazed and ceramic tiles for easy maintenance. There was no water tap or connection inside the toilet. A common water tap had provision for a small 4 litre bucket and a mug for body washing. Ash for sprinkling on faecal matter was stored in a bucket within the cubicle. This first ECCT had a block for men and another for women, each with seven cubicles, and mounted EcoSan pans for the physically challenged and senior citizens.

Figure 4.5

The Inside of a EcoSan Toilet

Based on the learnings from these experiences, a second ECCT with additional facilities and features was built at Parislathuri road. A system for disposing sanitary napkins and an incinerator was added in the women's block. A glass slit was added in the compost chamber for demonstration purposes, to observe dehydration. To prevent direct exposure to sunlight and crystal formation, the washing water and urine diversion pipes were given a higher gradient and were concealed. Handles were added to the lid of the drop hole.

Both ECCTs provided clean, environmentally friendly and cheap community toilet facilities along with additional benefits in the form of composted faeces and urine which could be used by farmers on their fields. In 2009, it was estimated that around 1,000 people (70% being women) used the toilet daily. The facility also provided opportunities for research related to using recycled human waste as organic nutrient for agriculture. After 6-12 months, excellent returns from agriculture by using the compost from the urine and the dried faecal matter prompted civic authorities to consider offering a payment to users of EcoSan public toilets. SCOPE, duly supported by state officials, introduced the first incentive of its kind in the world at the Saliyar Street ECCT, aptly calling it the 'Use and Get Paid Toilet' instead of 'Pay and Use Toilet.' This new year gift to the community was inaugurated on January 15, 2008, and meant paying toilet users 0.1 rupee per visit. This demonstrated the waste's economic value if properly recovered and used as a nutrient. Many users were a little embarrassed initially but everybody loved the idea.

Figure 4.6

Toilet with Ease of Access for Senior Citizens

Subsequently, a Memorandum of Understanding (MoU) was signed with the Tamil Nadu Agricultural University of Coimbatore to scientifically study the effects of using liquid urine and dehydrated composted human faeces on crops. Laboratory tests on dehydrated faeces have shown that pathogen and *E. coli* presence was nil, but more studies are being done to establish credible standards for safe agricultural practices. This approach can be broadened to cover all organic material generated by households, such as kitchen and food wastes, which can be sorted and composted into recyclable material rather than mixing with solid waste and dumping.

The Musiri town model has inspired several such efforts in India, mostly in peri-urban and rural areas. Adoption in the tsunami-affected eastern coastal areas is very significant. High saline water-tables in such areas have rendered conventional systems ecologically unsafe. In both public and private urban domains, the Musiri waterless toilet offers a totally new concept in designs for urban sewerage

Conclusion

There is a significant challenge for urban local planners, architects and builders to consider more economic and ecologically safe urban sanitation alternatives, such as introducing low-water or no-water human waste disposal systems, including vacuum flush toilets and separate piping systems for human waste collection. Promotion of such alternatives would significantly reduce contamination of fresh water while reducing the incidence of waterborne diseases caused by excessive use of water for transporting and disposing human waste.

Urban designers must take account of the spatial planning and engineering design implications of implementing innovative water supply and sanitation systems that will remain sustainable despite climate change, irrespective of their cost and irrespective of the fact whether the cities or towns are in the developed or developing worlds. Environmental costs and benefits must be incorporated into design. The time has come for urban planners, architects and engineers to create and share new designs—that are cost and resource effective, ecologically safe and environmentally sustainable—into the next millennium.

References

Chennai City website: *www.chennnai.tn.nic.in.*

Metrowater (Chennai) *website: www.chennaimetowater.tn.nic.in.*

5

JOSYULA LAKSHMI

Recharging Water Governance Systems

Introduction

The 'water crisis' of today is a reflection of its improper governance and lacunae in effective accountability mechanisms. The resource is being utilised without concern for its utility for future generations. It is said that 80 per cent of the people living in rural areas do not have access to safe drinking water; and five times more rural dwellers than city dwellers lack safe water.

Bettering the practices in water governance is the need of the hour and requires thought and deliberation. With multiple usages that water is put to, its importance as the 'lifeline for survival' has to be reiterated. Progressive legislations have opened up spaces for the public to demand participation in the decision-making and greater accountability in the state's water governance systems.

The present chapter diagnoses water governance issues and suggests possible steps to ensure its better governance. Some of the tools and mechanisms for better governance of the resource are suggested, including decentralisation, transparency mechanisms and equitable distribution. This chapter examines requirements from the vantage point of some of the new and proposed legislations in the country.

One of the targets of the Millenium Development Goal (MDG) number 7 is to: "Halve, by 2015, the proportion of people without sustainable access to safe drinking water and basic sanitation." The United Nations General Assembly (UNGA) adopted a similar draft resolution on the human right to water and sanitation by a recorded vote of 122 in favour and none against the motion. By doing so UNGA and its member country's committed themselves "to provide financial resources, build capacity and transfer technology, particularly to developing countries, in scaling up efforts to provide safe, clean, accessible and affordable drinking water and sanitation for all" (United Nations General Assembly, 2010). The internatioanl commitments showcase the significance accorded to the resource and the urgency of acting upon it. India is one of the 122 countries to have voted in favour of it at the UNGA in 2010.

In sync with the tempo of proactive affirmative actions, the development discourse and academic interest in the subject has evoked all round enthusiasm on the subject. Water is a finite resource, and it is feared to trigger the next world war (Chellany, 2011). The likelihood of the prediction does not seem too distant (Grey and Sadoff, 2007) with an estimated doubling and trebling demands for the resource and the competing pressure from its multiple usages [Asian Development Bank (ADB), n.d.] and the users' dependant on it. Decision-making about water resources and its democratic management is expected to enable its sustained existence and efficient usage.

Right to water is derived from the constitutional right to life. However, this right has not yet been formulated as one in the constitutional guaranteed rights in the country. The more recent draft National Water Policy 2011 (Ministry of Water Resources, 2012) put together for comments by the Ministry of Water Resources has also not spelt this right in as many words (Verghese, 2012). Yet this in no way need deter its administrators from strengthening the institutions and localising the resource management and introduce appropriate measures for its decentralised governance.

Right to water is for now extrapolated from the Supreme Court stand on the issue when it read the right to clean drinking water as a part of right to life (in the context of cleaning up Ganga and Yamuna). Right to drinking water is thus implied in the right to life. Denial of the same (in any form) is a violation of the basic right for survival. Availability and access to clean drinking water for the survival and well-being of human and animal life is thus critical.

Democratic Decentralisation

Democracies are defined as states 'of the people,' 'by the people' and 'for the people,' thereby reinforcing the shaping of the state through people's active participation. The reinforcement is given a further fillip, through the process of sharing the decision-making process with the other 'lower' levels of governance namely the federal and local levels of governance institutions. It is imperative that the constitutional division of subjects between the union, the states and the local bodies is devolved appropriately and their devolution well articulated. Ideally and practically so, these should be performed as per the principle of subsidiarity, that what can be best done at the lowest level must be delegated and designated for those to do.

Over the last two decades, decentralisation as an idea and practice gained prominance, because democracies have emerged to represent a powerful minority views and their decisions. As such, decentralisation is understood to ensure that decisions are taken at the level closer to reality and are hence broad based. The rationale for decentralisation refers to the devolution of the powers to the lowest of levels so as to broad base decision-making to make it seem to be reasonably a majority view.

The idea of democratic decentralisation and its actualisation got a fillip in India when the 73rd and 74th Amendment to the Constitution had been introduced in 1992. While the former accorded powers to the rural-level governance instititutions, the latter did so for the urban areas. The amendments are a watershed in the history of democratic decentralisation in the country. The 73rd Amendment reinforces the commitment of the governance institutions at the higher level to share decision making with and empower lower/rural local bodies in effectively managing the local affairs. As per the Eleventh Schedule (Article 243G), 29 subjects have been identified for rural local bodies to manage. Drinking water is one of these.

Armed with the progressive amendment, over the last two decades, efforts have been made to hand over powers to the local bodies. The process met with limited success and in few states.

States like Madhya Pradesh, West Bengal and Kerala (Planning Commission, n.d.) have taken the process of decntralisation forward by devolving the subjects and functionaries. Other states have had a faltering start and the process remains derailed in several other states (especially the PESA states). Experience of devolution of powers in various states on majority of the listed subjects is varied and disappointing to say the least. While a decentralised management of the resource (Johnson, 2003) like water would be ideal and desirable, alternative ways of cooption may be adopted for a well-informed citisenary to take social responsibility of the resource.

The Importance of Water Governance

Water is a key element that underpins lives and livelihoods, more so in rural India. It is an input to almost all rural productions in agriculture, industry, energy, and daily household usage and in management of the ecosystems. It thus has to be managed rationally and its multiple usages prioritised. Further its availability and access has to be distributed fairly among all the rural population, irrespective of the strata that they belong to—precisely because the needs are the same. Thus the idea is one of 'inclusive water governance' and thereby its management. Unmanaged water is also a cause of suffering and devastation, through seepages, floods, contamination and disease. That is a different issue and will not be dealt with in this paper.

Harnessing the productive potential of water and limiting its destructive impacts has been a constant struggle for society, and underscores the need for water security—defined as "the availability of an acceptable quantity and quality of water for health, livelihoods, ecosystem and production, coupled with an acceptable level of water-related risks to people, environments and economies" (Centre for Science and Environment, 2012).

Over the past years, discourses on water and its management have meandered from water rights to water and democracy. At the behest of bilatereal donors, it has moved from participatory hydrological management to water governance. The importance attached to the sector is not without reason. The concept of water democracy as discussed earlier has gained prominence because of some inherent problems. For instance, the resource is not distributed equally among all and there are several issues pertaining to its regulation, usage and management. While water management pertains to its day-to-day handling, water governance refers to the decision-making mechanisms and process that are in place. Good water governance calls for an effective and rational resource management and its just, fair and equal distribution. As resource, water is a common resource that calls for efficient handling, collective sharing and responsibly conserving it with equity as a value. It is obvious that it has not been done so far and hence the dire need to address it now. It could then be hoped that better water sense prevails.

It is widely held and correctly perhaps that the traditional systems of water harvesting were people friendly and rational with a keen eye on sustainability of the resource. It is for us to take lessons from the good practices of the past and ensuring that the resource does not become scarce.

Setting the Institutional Framework

An institutional framework for better water governance calls for the central and the federal states (that have been the controlling authroity on this resource so far) to delegate the management authority to local bodies and introduce mehcanisms for greater transparency and more accountability to ensure its sustenance in the long run. In respect of allocation of responsibilities between the state and centre on water in the Constitution, water is a matter included in Entry 17 of List-II which is the State List.

Today as we discuss about water governance issues, it is just as obvious now, that in the years to come, the majority population of the country that reside in rural areas will face a critical water stress. The rural water supply programme thus will face serious challenges toward meeting the expanding needs of a fast growing population, increasing demand for higher service levels accompanied by rapid depletion of fresh water availability. Water resource is stressed to such a point that demands on it are from multiple sectors and are on a steady rise. Stretching the resource beyond the seams is bound to affect the lives and livelihood of rural persons, resulting in mass migration to urban areas. This already is a reality as the resource is creating a rural-urban divide (Nagaraj, 2012). Water productivity in existing uses must, therefore, increase in response to the growing demand and the shrinking supply. Clearly, better water management must start at the resource level in order to provide effective and equitable balancing of existing uses with the needs of other sectors.

In water governance, we expect the decisions pertaining to water to be taken in a manner that the resource is delivered to the citizens with a sense of social resposibility—both in policy and in practice. From an individual point of view, individuals and groups influence service standards and policy implementation formulation through their economic and social transactions of the resource. The institutional context of decision making would involve understanding the levels of decision-making and devolution of powers through the 'three Fs' (Functions, Funds and Functionaries) to the different levels.

The nature and process of decision-making on water becomes an important differentiator of the quality and the sustanability of the resource. More often institutional processes can place new information before decision makers, enhance their awareness of the interests held by various groups, and expose discussion and relationships to public scrutiny.

What could good governance practices in the water sector be? Let us examine some indicative list.

1. *Systemic risk identification—Within and outside:* It would be important to recognise likely areas of corruption/gaps within the work system in rural water system (RWS). Potential areas of sabotage, vague and inconsistent internal monitoring and potent sources of faulty procurement may be identified and efforts made to plug the same. Similarly possible areas of influence from outside that could hinder

the smooth functioning of the system must be identified for timely mitigation.

2. *Strengthening linkages:* In the area of water governance, all the stakeholders—right from the higher authorities to the end-users—need to constructively engage to complement each other's actions towards water governance. This area is found inadequate, as the end-user has always been at a receiving end till now. Changing this equation to a more positive linkage would, in the long run, help to strengthen the ties and thereby facilitate better resource management, where imperative convergence with other line sectors at every level has to be explored.
3. *Transparency and access to information:* It is mandated now under Right to Information (RTI) Act (2005) and demanded that greater transparency exists in the existing institutional framework and greater access exists to critical information about the system. The more that information is provided proactively, it would help in faster de-bunking of the so-called technical information. Making quick analysis of the critical quantitaive data, along with its qualitative ramifications for the people would go a long way in improving the quality of the output and its availability and access. In times of exigencies like natural disasters, the information thus provided would help curb health crises and reduce the spread and scale of epidemics due to drinking water contamination.
4. *Leveraging accountability:* Every positive step taken to improve the accountability mechanisms towards the end-users within RWS would be a potent area of transparency and good water governance. This could include making the system work proactively rather than reactively, introducing and mandating mechanisms for social control through social accountability tools like, social audits, community score card, citizen report cards and periodic feedback mechanisms coupled with third party evaluations for quality control.

5. *Improving service delivery:* In the present context of talks about privatisation of water in the country, it would be better idea to learn from painful experiences of the other countries that have adopted it. Experience shows that while privatisation helps the state to divest itself from the reponsibility of providing clean drinking water, it makes no qualitative change for the consumer (end-user) of this service. If at all, it would burden the citizen with an additional demand on the pocket for an inferior good affecting the physical and financial health. Mechanisms of better service delivery may be introduced within the existing service delivery system by stipulating standards and quality benchmarks at different levels of governance.

   This may soon be mandated as well, if the Citizen's Right to Grievance Redress (CRGR) bill is enacted.

6. *Periodic updation of data:* The fun of data analysis lies in its authentic collection and periodic updation. It then becomes amenable to analysis and facilitates accurate water governance. During normal times, data collection may be done less periodically, but during emergencies, it helps to have frequent updation of water data. At any point of time, the data should be available for analysis and not found wanting. An internal step in this direction is to ensure an appreciation for authentic data collection within the departments dealing with RWS.

Further to the governance areas identified above, this analysis would now look into implications for water governance with specific reference to the existing RTI Act, 2005, and the proposed Citizen's Right to Grievance Redress (CRGR) bill, 2011. While the RTI Act obliges the public authorities to part with information held by them, the latter bill outlines the responsibilities of the government departments towards citizens, and how someone who is denied the service can take action. Let us examine these in the order of their emergence.

Right to Information and Water Governance

In the wake of the RTI Act, 2005, it is expected that information held by the public authority is made available on demand or more conveniently for all, to be disclosed proactively in anticipation of possible queries. The RTI Act defines information as follows [Sec. 2 (j) (iii)]: "Any material in any form, including records, documents, memos, e-mails, opinions, advices, press releases, circulars, orders, logbooks, contracts, reports, papers, samples, models, data material held in any electronic form and information relating to any private body which can be accessed by a public authority under any other law for the time being in force."

There are two ways by which public authorities give information to a citizen under the RTI Act: (1) when demanded i.e., reactively and (2) by supplying even though it is not specifically asked for i.e., proactively. While the former mode of furnishing information has been the preferred mode of sharing so far, experience and good sense of governance says that the latter is a more desirable mode. Proactively furnished information is less likely to be asked for, thereby economising a citizen and a public servant's time, resources and energy in accessing it.

RTI Act entails that all forms of information—data, reports, files, manuals, guidelines, etc.—are disclosed to the citizens on demand. In departments that are responsible for such resources as the water resources, etc., the information available may be technical (quantitative) in nature and thus deemed to be above the comprehension of a lay person. It becomes helpful then to make it less technical and therefore intelligible to a citizen, for whom the department caters and is its *raison d'être*. It is a common practice to seek information that is electronic or paper-based under RTI. Guidelines seeking this type of information have been streamlined and there are standard procedures available for public use.

Sample as 'information' in the definition of RTI Act, however, is peculiar in that it is one information that needs further testing to qualify it as an intelligible piece of information. Such information could be perceived to include the quality of material used in works, material stocks procured by government or government-approved executioners (contractors/builders). In fact right now, the process of bidding, awarding the projects, information notations on public files and electronic information such as emails or procurement/tendering processes are open to scrutiny under RTI by the citizens. Citizens not only have the right to seek information related to material used, but also seek 'samples' as information under RTI Act. For this purposes, it is critical to define 'samples' as included in the definition of information in RTI act.

A sample as information under RTI act would include getting and examining 'a small part of' the following:

- materials used in construction,
- consumables,
- stocks (quality and quantity),
- fixtures and fittings used in executing of public projects, and
- procurement process and physical samples that contain qualitative information about the material that is planned to put to use or is put to use in the execution of public projects.

Sample invariably has to be tangible that could be seen, felt and its quality tested for genuineness and as per assurances made of its usage and long-term sustenance.

By expanding 'sample' and defining in terms of what it entails in the context of water governance, citizens could be assured of the quality of the consumable, the fixtures in use (used) and the construction material. Information on the quality and type of material that will be used for construction or execution of public projects is usually included in the technical proposals. Similarly, the specifications, standard of fixtures, specifications of consumables, etc., are included in detailed project proposals or technical reports of projects. It is common knowledge that there are glaring gaps in the quality included in the technical proposals on paper and the actual material used during execution of projects. It is uncommon, however, that the sample material is tested and audited publicly and action taken upon deviations found therein.

The information sought on the 'water quality' could be both the recordings in the proposals, files as well as actual water flowing in the pipeline. It so happens that while the source may be dispensing quality product, the resource may get contaminated in transit or may get contaminated at the end point. The procedure seeking information of the first kind (quality at source) is well documented and is likely to be readily available. It is at the transit and end point that samples (information) needs to be streamlined. There is a need to develop guidelines and/or toolkit that will simplify seeking information on technical quality or seeking physical sample itself and assessing the quality of the sample as defined against standard specifications.

Quality control and its monitoring is an essential process to ensure a high level of quality in a facility, product or service. The basic goal of quality control is to ensure that the facility, products, services and processes adopted shall meet specific/agreed standards and requirements in their ultimate deliverance. All public utility services including infrastructure development projects, such as water supply schemes for drinking and other purposes, sanitary works, pollution control and other services, need to be provided with acceptable quality and to the best satisfaction of various stakeholders and the end-users.

The quality control teams ensure that the product or service meets the specified standards of quality. If deficiency in detected in quality or problems are identified, the same have to be rectified immediately. Wherever required, the work will have to be temporarily halted till the rectification is complete. Citizens expect and can now draw information under RTI for public authority's adherance to quality procedures and products by way of demanding samples at different stages of execution of the project. Further, new projects normally declare to maintain highest standards of quality, safety and environmental protection at every stage namely: the formulation stage, designs and estimation stage, procurement of execution and quality control agencies, construction with quality, and operation and maintenance (O&M). The same may be audited for a reality check.

The RTI Act would not be effective if its nuances are not deciphered and made intelligible to all. An emphatic and effective proactive dislcosure of water source and supply mapping, quantitative availability, quality standards and public display of third party evaluation and consumer water confidence reports (whereby consumers could vouch for the quality of potable water in their locality at various intervals of time) may be collected and made available at periodic intervals. This initiative would deem to be a measure towards better water governance.

The institutions responsible for these may like to prepare and keep manuals, procedures and/or toolkits available with them that lay down step-by-step process for seeking 'sample information.' This could facilitate in making public project execution more transparent. The need to develop such a manual/toolkit assumes paramount significance, given the scope and magnitude for process-malpractices in material procurement and existing gaps/mismatch in the quality proposed on paper and the actual quality utilised in execution of the project.

From a citizen's perspective, due to the perceived technicalities in understanding 'samples,' governance advocacy groups and common citizens often make do with sub-standard products. The Consumer Protection Act (1986) is the only other recourse. But that often comes at the end-user stage and is more for grievance redressal and problem mitigation. Accessing information at the source and during the process (in transit) would be more meaningful in holding public authorities accountable, rather than waiting for the end-user feedback. As the adage goes, prevention is better than cure. Early detection at an early stage is thus better than a final stage damage control. In case of water, nipping it in the bud would help in making the process more valuable.

Information mismatch: In case of procurement of bulk stocks and laying down of piped water supply and sewerage lines, the information would basically be available on paper—the lay out, the design, estimates, etc. The laying out and actual on the ground situation would give an idea about its field level execution. Water that is envisaged to be flowing from those pipelines becomes the sample as its quality at source is tried to match with its end-user quality for purported compromises, if any.

Quality mismatch: When there is a mismatch between the permissible level of chemical contamination in water and its sample, there would be a need to identify the differentiator, single it out and separate it to ensure its source quality.

Seeking and providing sample as information will have two distinct stages. In the first stage, the information furnished by the contracting agency on the technical proposals related to the quality of products will be put to use and the standards that will be adhered are spelt out. This is usually paper-based information.

The second stage of sample information involves the 'material sample' itself. Procuring this sample from an ongoing project or completed project involves, collection of the sample (intrusive or non-intrusive), assessing the sample for quality either on the site or at specialised quality assessment laboratories and keeping a vigil on the quantity used.

A technical guide as to what constitutes a sample, how to seek 'sample information,' from public authorities in RWS and possible conclusions to be drawn from the sample could be a major contributor to the knowledge based on the various constituents of RWS.

Grievance Redressal and Accountability

Proposed Citizen's Right to Grievance Redress (2011) Bill

The proposed CRGR bill has as its objective the following:

1. Every public authority or government department to publish a citizen's charter listing all the services that the department has to render along with timelines.
2. A grievance redressal mechanism for non-compliance of citizen's charter will be in place.

The proposed bill provides for every public authority to appoint or designate Grievance Redress Officers (GROs) whose contact information will be shared with the public.

The GRO will provide the public with all necessary assistance in filing complaints. Within two days of the complaint being registered, the citizen who has filed a complaint will receive the response through short messages or email. A unique complaint number will be set up along with a time frame within which the complaint will be handled. The time frame cannot exceed 30 days from date of the complaint received. If citizens are unhappy with the decision of the central or state commissions, they can appeal to the Lokayukta or Lokpal.

As this is a proposed bill tabled in Parliament, we are yet to gather experiences of its execution. Several states like Madhya Pradesh, Goa and Bihar however have enacted their state legislations on Public Service Guarantees for Citizens. Their experiences show that citizens are more than willing to hold the government to account on all its public works. Agaisnt this backdrop, the citizen is truly empowered to question the government and its instruments.

Ensuring that all public services are made available to the citizens in a time bound manner would help to aim for service levels pitched at high level of excellance rather than mediocre. Professionalism and capable staffing would be the key to do that. With the goal of service excellance, one may combine merits of time-bound service delivery that may be incentivsed—institutionally and individually.

Social Accountability Tools

There is ominous urgency to formally introduce locally practicable tools of governance and accountability to make the processes more decentralised and thus manageable. Efforts to minimise dependance on the state level instruments as far as technical information is concerned must be encouraged and patronised. Besides building a cadre of village level technical animators (VLTAs), some social accountability tools have to be introduced at the field level and mandated to be followed up by the department and the line staff scrupulously. These are advisable to keep the citizens 'informed' about the ground water availability, its depleting nature and to plan alternatives locally.

Some of the social accountability measures that could be taken up are:

- Social Audit of Drinking Water:
  - *Water Resource (Availability) Audit (WRA):* That the resource is available in its fulness i.e., the minimum madated (40 lpcd) and its rational usage within a village may be ensured through the resource audit. The duration and timeliness of its availability may be audited. Its rational usage by all as against wastefulness may be gathered during the two mandated *gram sabhas* and special *gram sabhas*, making it a regular routine event. WRA could also take stock of the source to ensure its availability for the next few months/years. Based on the availability of the resource, the villagers may plan its prudent regulation.
  - *Water Quality Audit (WQA):* Sample kits and tools and their usage to check the quality of available drinking water would come a long way in ensuring that local contaminations are curbed at the source itself. Sample testing techniques and conclusions that can be drawn from these could be made available to the local villagers.
  - *Water Equity and Inclusiveness Audit (WEIA):* An index to check the fact that water is equitably distributed among all the inhabitants of a village and that its coverge is fool proof as far as the social strata of that village is concerned would help in gauging its social coverage.
- Community Score Card: This tool could be used to engage the service seekers (villagers) in an interface with the service providers (officials) whereby at a given point of time and periodically thereafter, they rank each other on the basis of a few key identified indicators of service level performance. There could be several rounds of score card interfaces.

- *First round:* At the initial stage when both the parties identify indicators for the better governance of available local water and mark their preferred score on a 10-point scale. The groups' scores are averaged to arrive at a citizen's average scoring on each indicator and another for the service provider. A round of negoitiation on the issue with both parties analysing the situation would ensue. The process is continued and a consensus is arrived on each indicator with an action point to be planned from identified gaps and responsibilities assigned to local office bearers and citizens against a rational agreed time line.
- *Second Round:* After a gap of a month or two, the above process is repeated. New action points emerge for both parties and new timelines are set for the same.
- *Third Round:* Repeat of actions under the Second Round.
- *Fourth round:* Interface meetings to conclude the levels of decision-making should they be beyond the local parties.

- Public Expenditure Tracking: For every penny spent with the end-user of water in mind, the qualitative and quantitative outcomes for the end-user must be similar to the source level user. This would ensure that the money is spent for a worthy cause and would result in the desirable outcomes. Public audit of the financial resources being expended to ensure the outcome augurs well for transparency and better accountability and thereby better governance.

The above-mentioned tools are but few social accountability instruments for achieving the desirable outcomes for water literate citizens. These are not exhaustive and could be further explored.

Conclusion

Making water governance thus perceivably democratic through various modes like realising decentralisation of decision-making, delegating powers to Panchayati Raj Institutions (PRIs) and the Community groups, would augur well for transparency and accountability. This calls for systematic application of social accountability mechanisms already proposed RTI Act and the proposed CRGR bill. Foresight and a vision in this direction would even out possible areas of friction between governance institutions and the end-users of the resource. Good water governance will require interventions at mutiple levels, collaboration between multiple actors and convergence of multiple levels of actions.

References

Asian Development Bank (n.d.). *Water Resources Development in India: Critical Issues and Strategic Options*.

Centre for Science and Environment (2012). "Dying Wisdom: Rise and Fall of the Traditional Water Harvesting Systems", *State of India's Environment—Report* (1997).

Chellany, Brahma (2011). *Water—Asia's New Battleground*. Washington, DC: GU Press.

Grey, David and Claudia W. Sadoff (2007). "Sink or Swim? Water Security for Growth and Development", *Water Policy* 9(6): 545–71. IWA Publishing.

Johnson, Craig (2003). *Decentarlisation in India: Poverty, Politics and Panchayati Raj*. London: ODI.

Ministry of Water Resources (2012). *Draft National Water Policy*. New Delhi: Government of India.

Nagaraj, G. (2012). *Water Crisis Triggers Urban-rural Divide*, *The Hindu*, March 10.

Planning Commission (n.d.). *Decentralisation and Panchayati Raj Institutions* (*PRIs*). Chapter 10. New Delhi: Planning Commission.

United Nations General Assembly (2010). *GA/10967: Proceedings of UN General Assembly*. July 28.

Verghese, Shiney (2012). "Turning off the Tap on Water as a Human Right", *The Hindu*, February 16.

6

HEMANT JOSHI and SUNEET SETHI

The Rural Water Sector in Rajasthan

Introduction

Rajasthan is blessed with the Aravallis—one of the oldest mountain ranges of the world and the golden sand dunes of the Great Indian Desert—the only desert of the sub-continent. With an area of 342,239 sq km, it is the largest state of the country having 10.41 per cent of the country's area, 18.70 per cent livestock of the country and 5.5 per cent of nation's population; it has low water resources with only 1.16 per cent surface water and 1.72 per cent groundwater of the country. Sixty-six per cent area of the state is desert. Drought is a recurrent phenomenon in the state.

The state has extreme climatic and geographical conditions. The Aravalli range of hills divides the state in two parts. The north-west region has scanty and uncertain rainfall while south-east region has fair to good rainfall. The state has water problem in terms of both quantity and quality. In most parts of the state, groundwater is either saline or has excess fluoride. Groundwater is the major source of drinking water and over 91 per cent of the drinking water demand is met by groundwater. About 32,000 habitations have water quality problem. About 60 per cent of scheduled caste (SC) habitations lack an independent source and about 90 per cent are located in desert districts. Out of 249 blocks, 198 are under category of overexploitation due to depleting potable groundwater reserves and deteriorating groundwater quality. Reliable surface water sources i.e., Indira Gandhi Canal and the Chambal and Mahi rivers are located on borders of the state requiring huge cost outlays for transferring water to interior areas.

Water has become a critical and scarce natural resource. The proper utilisation of water resources is necessary to ensure availability of water in right quantity and quality for diverse purposes. In recent times, on account of the increase in population, urbanisation, industrialisation and the use of chemicals in agriculture, there is an ever-increasing threat to surface water and groundwater quality. Concern for water quality has thus become an important matter and calls for urgent attention.

The paper discusses the major challenges in the field of decentralised governance in rural water sector in the state. It delineates the steps already taken up and proposed in near future for achieving complete decentralised governance in rural water sector.

Present Scenario in Rajasthan

Overall rural drinking water scenario in Rajasthan is shown in Table 6.1.

Table 6.1

Rural Drinking Water Scenario in Rajasthan

Particulars	*No. of Habitations with Population Coverage (as on 20.2.2012)*			*Total Habitations*
	>=0 and < 25%	*>=25 and < 100%*	*100%*	
All habitations	32979	19188	68966	121133
SC dominated habitations	4296	2091	8934	15321
ST dominated habitations	2855	3578	16487	22920
Minority dominated habitations	5093	1114	5483	11690

Source: PHED (n.d.). *Pragati Vivranika 2011-12*. Rajasthan.

Twenty-eight per cent of SC dominated habitations have less than 25 per cent drinking water coverage and 58 per cent have 100 per cent coverage. Twelve per cent of ST dominated habitations have less than 25 per cent coverage, while 72 per cent have 100 per cent coverage. On the other hand, 43 per cent of minority dominated habitations have coverage less than 25 per cent and 47 per cent have 100 per cent coverage.

All of the 222 towns, 37,603 villages and 49,869 other habitations of the state have been provided with drinking water. While this achievement with regard to provision of drinking water has been impressive, the bacteriological and chemical pollution of drinking water sources pose serious threat to public health. The tremendous use of groundwater for irrigation and other purposes has resulted in sharp decline in the groundwater table and changes in the natural chemistry of groundwater. Natural contaminants such as fluoride, nitrate, iron and salts are increasing in groundwater in levels which make it unfit for drinking and pose a risk to health as shown in Figure 6.1.

Figure 6.1

Quality Affected Habitation Contamination-wise as on 01/04/2011

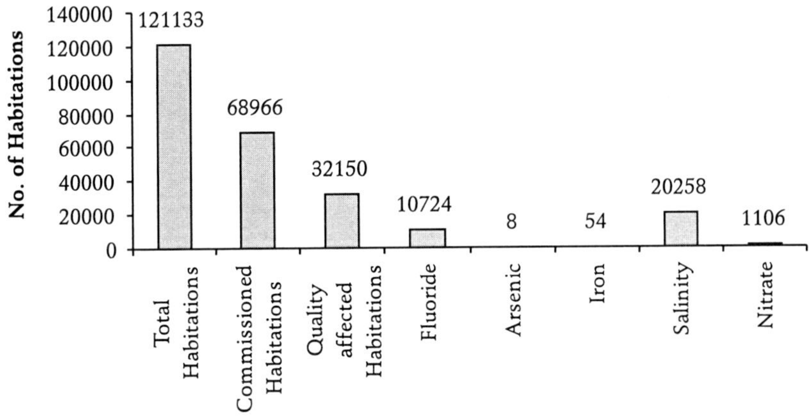

Source: Census of India, Government of India, 2011.

It is observed that out of total 68,966 commissioned habitations 32,150 villages are quality affected which is around 47 per cent of total covered villages.

This situation has led to a phenomena euphemistically described as slip back habitations. Due to this situation, the commissioned habitations are now only 57 per cent and 43 per cent habitations are to be covered fully as shown in Figure 6.2.

Figure 6.2

Status of Habitation Coverage

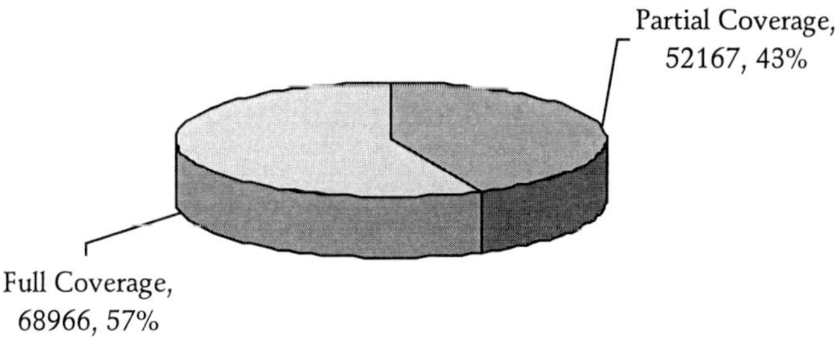

Source: PHED (n.d.). *Pragati Vivranika 2011-12*. Rajasthan.

As per the new National Rural Drinking Water Programme (NRDWP) guidelines a habitation cannot be counted as commissioned so long as the water supplied to the community is not potable. The present stage of groundwater development in the state is 125 per cent.

Decentralised Governance

Decentralisation is the policy of delegating decision-making authority down to the lower levels in a set-up relatively away from a central authority. A decentralised set-up shows fewer tiers, wider span of control, and a bottom-to-top flow of decision-making and flow of ideas.

Decentralised governance is the transfer of responsibility for the planning, financing and management of public functions from the central government or regional governments and its agencies to local governments i.e., *gram panchayats* (GPs). The three major forms of administrative decentralisation—deconcentration, delegation and devolution—have different characteristics. In this case, devolution is the form of decentralisation which has been discussed in this paper.

Devolution

Devolution is an administrative type of decentralisation. When governments devolve functions, they transfer authority for decision-making, finance and management to quasi-autonomous units of local government i.e., GPs. Devolution usually transfers responsibilities for services to local governments that elect their own elected functionaries and councils, raise their own revenues, and have independent authority to make investment decisions. In a devolved system, local governments have clear and legally recognised geographical boundaries over which they exercise authority and within which they perform public functions. Administrative decentralisation always underlies most cases of political decentralisation.

The 73rd constitutional amendment of India says that the state or provincial governments should transfer the responsibility of drinking water supply, particularly the operation and maintenance (O&M), to the PRIs (panchayati raj Institution-village government) through an appropriate state level rural water supply policy using an integrated approach to facilitate water sector reform. With the 73rd Constitutional Amendment Act of 1992, fully representative local government structures came into effect at three levels for rural areas. The important features of the 73rd Constitutional Amendment are:

- constitution of *gram sabha* at the village level consisting of all the electors,
- three levels of directly elected local self-governments for rural areas,
- regular elections at five-year intervals,
- reservation of seats for SCs and STs in proportion to their population,
- reservation of one-third of the seats for women, including those of offices of chairpersons at all levels,
- constitution of state finance commissions for considering the financial requirements of the local bodies and making appropriate recommendations to the state governments, and

- establishment of state election commissions for the conduct of local body elections. Ensuring maximum democracy and maximum devolution,' the 73rd Amendment was considered as a major watershed with respect to local self-government institutions.

The Status of Decentralised Governance in Rajasthan

Mahatma Gandhi nurtured the vision of *gram swaraj*. His dreams were eventually fulfilled when democratic powers were decentralised and the local government-system in rural areas was established in the form of GPs throughout the country. Rajasthan was one of the major states where the three tiers panchayati raj system were introduced on the recommendations of Balwant Raj Mehta Committee. This was inaugurated by Pandit Jawahar Lal Nehru on October 2, 1959, at Nagaur. Chaudhari Kumbharam Arya (1914-1995) a famous freedom fighter, famous Agrarian leader and parliamentarian was the founder member of the Panchayat Raj system in Rajasthan. However the *panchayat* system had no constitutional status until 1993. On April 23, the panchayati raj gained constitutional status through the Constitution (73rd Amendment) Act, 1992. The main objective of Panchayati Raj is to develop a system of devolvement and democratic decentralisation of powers with an aim of fostering rapid social and economic development and prompt justice. Some social developmental measures include:

- Improving school infrastructure and quality of education.
- Increase water and electricity availability.
- Reduce the occurrence of communicable diseases.
- Increase income generated per household.

The Directive Principle of the State Policy, Article 40 of the Constitution of India, provides that the state shall undertake steps to organise village *panchayats* and these *panchayats* will be empowered with necessary authority to enable them to function as units of self-government.

The Panchayati Raj system of Rajasthan and other states consists of a three-tier structure:

- The village *panchayats*.
- The *panchayat samities*.
- The *zilla parishads* (ZPs).

The Rajasthan *panchayats*, *panchayat samities* and ZPs were assigned 33, 58 and 19 subjects respectively by the Rajasthan Panchayat Raj Act (1994), however these were subject to conditions laid down by the government. The 33 districts of the state are divided into 249 *panchayat samities* and 9,177 village *panchayats*. Currently elections to the Rajasthan *panchayats* are frequently held and are supervised by the state election commission. In 2000, more than 1.2 lakh representatives were elected to the *panchayats*.

In keeping with the spirit of the 73rd amendment, the Rajasthan government has transferred the administrative controls of five departments with funds, functions and functionaries to *panchayat department* on October 2, 2010. The departments are:

1. Primary Education.
2. Health.
3. Women and Child.
4. Social Welfare and Justice.
5. Agriculture.

The Total Sanitation Campaign was also transferred to Panchayat Raj Department.

At present, Janta Jal Yojnas are being maintained by *panchayats*. In these schemes, a pump is installed on the open well and the village gets water from this well through this pump. The payment of electricity bill is made by *panchayats*. An amount of ₹ 500 is paid to the operator who runs the pump. The repair of the pump is also done through *panchayats* if it goes out of order. The GPs are also taking up the work of construction of hand pumps through funds received under the State Finance Commission (SFC) and Thirteenth Finance Commission (TFC).

Presently the repair of hand pumps is officially with the *panchayats* in the sense that administrative control of hand pump repair *mistries* is with Panchayats Raj Department, but the funds and functionaries have not been transferred. The Panchayat Raj Department has to manage the repair work with their existing staff. There are 3 lac hand pumps in the state. The Panchayat Raj Department finds it difficult to repair these hand pumps, therefore the Public Health and Engineering Department (PHED) has to undertake the work of repair of hand pumps almost nine months of a year as part of a special hand pump repair campaign.

The rural drinking water schemes have still not been transferred to *panchayats* in the true sense of the word. The work of repair of hand pumps was transferred to *panchayats* but the transfer is not wholesale therefore there are problems encountered in this work.

In addition to the above, the state government has also decided that projects costing more than ₹ 25 crore within the village will be maintained by a five-member committee selected in *gram sabhas*. There will be one committee of five members in each village. This committee will appoint an operator in each village who will assist the committee in O&M. The system from the source to transmission pipe lines up to the entrance of the village will be maintained by PHED, but inside the village the water will be provided through a bulk meter. The five-member committee will take care of operation and maintenance as well as revenue collection. In addition to the village water and health committee in each village, there is a 15 member village water, sanitation and health committee at every GP level. This committee consists of members of the fifth committee of GP for social welfare and all the grass-roots level functionaries along with representatives of SC/ST and women groups. The GP committee has all the administrative, financial and technical powers to undertake O&M work. This model is already running successfully in Apni Yojana Churu district in 397 villages. This model is being replicated in all the major projects of the state.

There are other types of water supply systems also in the state. They are as follows:

- Pump and tank schemes.

- Regional water supply schemes.
- Piped water supply schemes.

These schemes continue to be maintained by PHED with the help of a strong force of 1,660 qualified and experienced engineers. The above schemes are planned, designed and implemented also by PHED. The schemes are designed on the basis of departmental norm of existing population. The pump and tank (P&T) and piped schemes are designed in a village if the existing population is more than 1,500 and 4,000 respectively. At present there are still 572 villages in the state having population more than 4,000 but still not having piped water scheme. Total number of benefited villages and habitations through various rural water supply schemes are shown in Figure 6.3.

Figure 6.3

Number of Benefited Villages and Habitations through Various Rural Water Supply Schemes

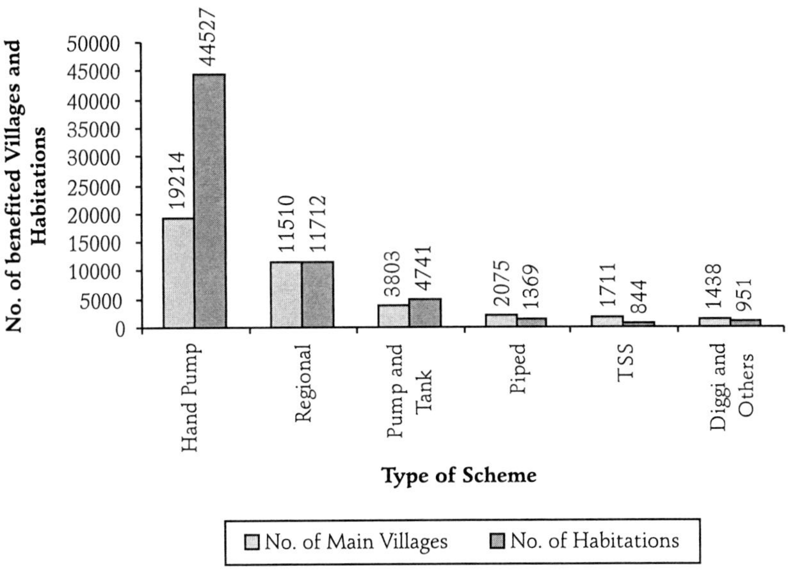

Source: PHED (n.d.). *Pragati Vivranika 2011-12*. Rajasthan.

Percentage-wise number of benefited main villages is shown in Figure 6.4.

Figure 6.4

Number of Benefited Main Villages through Various Rural Water Supply Schemes

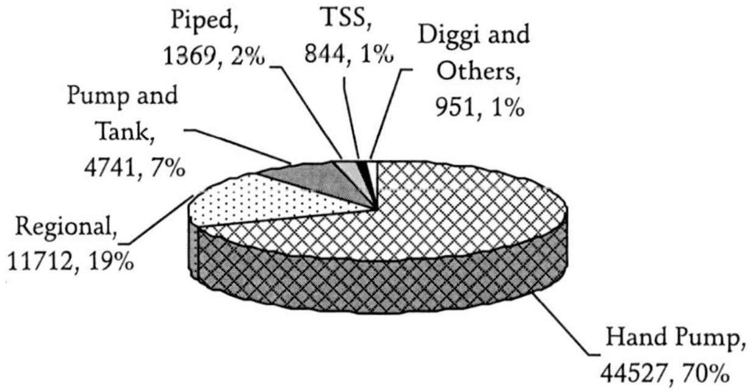

Source: PHED (n.d.). *Pragati Vivranika 2011-12.* Rajasthan.

Figure 6.5 illustrates year-wise hand pump repairing campaigns during 2001-2011.

Figure 6.5

Number of Repaired Hand Pumps during the Hand Pump Repairing Campaign (Year 2001-2011)

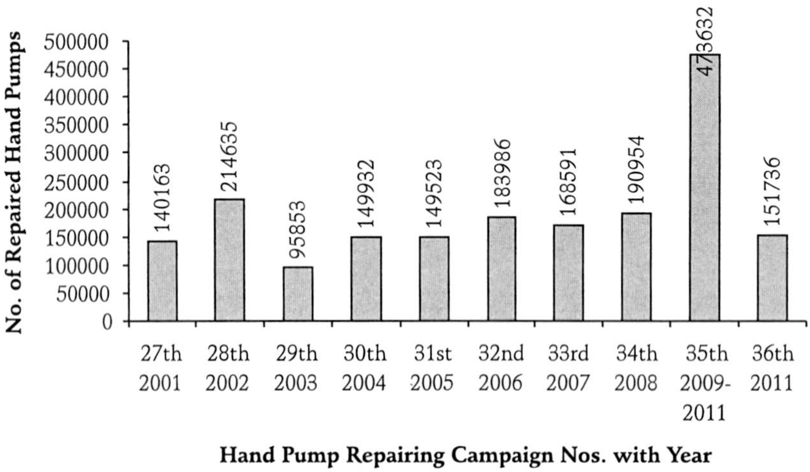

Source: PHED (n.d.). *Pragati Vivranika 2011-12.* Rajasthan.

It is clear from the figure that from the year 2001 to year 2011, hand pump repairing is ranging from 38 per cent to 72 per cent. The revenue is not even 1/10th of the expenditure.

Issues and Challenges in Decentralised Governance in Rajasthan

- The decision to transfer drinking water to *panchayats* is a political one. There was a lack of confidence at the political level in terms of capacity of Panchayat Raj Department to successfully maintain these schemes. But now with the transfer of five departments to PRIs, the ice has been broken and there are chances that other subjects in the 11th schedule may also be gradually given to PRIs.
- The mind set of PHED engineers to continue with the *status quo* is also playing a role in postponing the decision.
- The lack of capacity of PRIs to maintain the water supply schemes needs to be increased through training.
- Potable groundwater reserves are rapidly being depleted, while groundwater quality is deteriorating.
- The conflict of interest of different regions and different user categories in implementation of surface-source-based schemes and allocation of water needs to be overcome.
- The state faces a major water crisis every third or fourth year due to recurrent drought. The entire developmental programming revolves around drinking water in the state. Therefore, the state government is cautious and want to move forward slowly and steadily without causing any consternation in the masses on account of transfer of drinking water sector to PRIs.

Management Devolution Index

The Government of India (GoI) has devised Management Devolution Index (MDI) which on the basis of certain indicators finds out the extent of devolution of power to the PRIs. The devolution index is divided in four categories which are as follows:

- Transfer of functions to PRIs.
- Availability/transfer of funds to PRIs.
- Functionaries made available to support PRIs.
- Good management indicators.

The score of Rajasthan in this index is not very impressive. The score of Rajasthan is only 17 out of 100. If a comparison is made with other states, then the ranking is 12 out of 28 which is not very bad. The detail is given in Table 6.2.

Table 6.2

State-wise Calculation of Management Development Index

S. No.	*Name of State*	*Score*
1	Karnataka	46.4
2	Tamil Nadu	41.7
3	Andhra Pradesh	31.8
4	Gujarat	28.1
5	Maharashtra	24.7
6	Uttar Pradesh	24
7	Punjab	23.8
8	Uttranchal	23
9	Kerala	22.1
10	Chhattisgarh	20.4
11	Jharkhand	18.7
12	Rajasthan	17
13	Sikkim	16.4
14	Himachal Pradesh	16.3
15	Madhya Pradesh	16.3
16	Tripura	15.3
17	West Bengal	13.2
18	Nagaland	12.8
19	Haryana	12.7
20	Mizoram	9.5
21	Bihar	9.4
22	Assam	9.1
23	Odisha	8.8
24	Arunachal Pradesh	8.2
25	Meghalaya	5.3
26	Goa	0
27	J&K	0
28	Manipur	0

The performance of Rajasthan in various indicators is as given in Table 6.3.

Table 6.3

Calculation of Management Development Index

S. No.	*Devolution Indicators*	*Weight*	*Rajasthan*		
			Claimed	*Awarded*	*Reason*
1	**Transfer of functions to PRIs**				
1.1	The State Acts and/or executive orders/MoUs clearly defining the transfer of responsibility for infrastructure creation of hand pumps to PRIs.	2	Yes	2	
1.2	The State Acts and/or executive orders/MoUs clearly defining the transfer of responsibility for infrastructure creation of Single Village Piped Water Schemes to PRIs.	2	Yes	0	No specific documents
1.3	The State Acts and/or executive orders clearly defining the transfer of responsibility for O&M of hand pumps to PRIs.	2	Yes	2	
1.4	The State Acts and/or executive orders clearly defining the transfer of responsibility for O&M of Single Village Piped Water Supply Schemes to PRIs.	2	Yes	2	
1.5	Proportion of hand pumps whose O&M has been transferred to PRIs.	2		2	
1.6	Proportion of Single Village Piped Water Schemes whose O&M has been transferred to PRIs.	2		0.6	
1.7	The VWSCs should be standing/sub-committees of GPs under the State Act/Rules.	2	Yes	2	
1.8	Proportion of drinking water sources for which water quality testing done during the previous years as per the IMIS.	6		1.5	
	Total for functions	20		12.1	

contd...

...*contd*...

2	**Availability/transfer of funds to PRIs**			
2.1	Proportion of NRDWP (Coverage and Quality) funds (Central + State share), transferred to PRI*/DWSM subordinate to ZP accounts.	15	No	0
2.2	Proportion of NRDWP (O&M) (Central + State share) transferred to GPs accounts.	15	No	0
2.3	Whether untied grants from State Plan/non-plan have been transferred to all GPs in preceding financial year.	5	No	0
2.4	Whether untied grants from State Finance Commission have been transferred to all GPs in preceding financial year.	5	No	0
2.5	Whether unit charges of electricity for pumping in by PRIs are equal to or lesser than lowest slab of unit charge for domestic consumers.	5	No	0
2.6	Percentage of water charges demand collected by PRIs.	5		0
	Total for funds transfer to PRIs.	50		0.0
3	**Functionaries made available to support PRIs**			
3.1	Proportion of blocks where block level Diploma/Graduate engineers are available in PHED/PRED/ZP/BP @ one for 1 lakh rural population exclusively for rural water supply and sanitation.	5		0
3.2	Proportion of filled up DWSM Consultants positions.	5		2.5
3.3	Proportion of filled up BRC Coordinators positions.	5		0
3.4	Proportion of VWSC members trained in RWS functions for at least two days in the previous year.	5		0
3.5	Proportion of NRDWP Support funds spent on IEC and HRD activities.	5		2.1
	Total for functionaries support to PRIs	**25**		**4.6**

.*contd*...

...contd...

4	**Good Management indicators**		
4.1	Proportion of households in State provided with household connections	2	0
4.2	Proportion of villages with bulk water supplies metered	3	0.3
	Total for good management Indicators	**5**	**0.3**
	Grand Total	**100**	**17.0**

The consolidated position of the state in four categories is shown in Table 6.4.

Table 6.4

Consolidated Position of the State in Management Development Index

S.No.	*Indicator*	*Weightage*	*Awarded Points*	*Remarks*
1	Transfer of functions to PRIs	20	12.1	There are some indicators which have not been reported correctly on the national website of DDWS due to lack of reporting or wrong reporting
2	Availability/transfer of funds to PRIs	50	0	
3	Functionaries made available to support PRIs	25	4.6	
4	Good Management indicators	5	0.3	
	Total	100	17	

The indicators where Rajasthan has done well have not been reported properly on the website. They are as follows:

- Whether untied grants from state plan/non-plan have been transferred to all GPs in preceding financial year. *Status:* In year 2011-12, ₹ 45 crores has been transferred and in year 2012-13, ₹ 96 crores have been transferred.
- Whether untied grants from State Finance Commission have been transferred to all GPs in preceding financial year. *Status:* In year 2011-12, ₹ 700 crores transferred.
- Percentage of water charges demand collected by PRIs. *Status:* 100 per cent in case of schemes covered under major projects.

- Proportion of blocks where block level diploma/graduate engineers are available in PHED/PRED/ZP/BP @ one for one lakh rural population exclusively for rural water supply and sanitation.
  Status: 1,660 engineers are available. In every district, 2-3 executive engineers are looking after the work of RWS; under them 10-15 A.Ens and 20-30 J.Ens are assisting in the work.
- Proportion of filled up block resource centre (BRC) coordinators positions.
  Status: BRCs have been created at every block in the state to provide continuous support in terms of awareness generation, motivation, mobilisation, training and hand holding of village community, GPs and VWSCs. There are two to four functionaries in the BRC depending on the population of the block. In the state, 249 block coordinators and 667 cluster coordinators are to be appointed; out of which till date, 134 block coordinators and 338 cluster coordinators have been appointed. A total of 472 coordinators have been appointed out of 916, i.e., 52 per cent.
- Proportion of VWSC members trained in RWS functions for at least two days in the previous year.
  Status: Total 63,500 VWSC members have been trained uptill now. The target for VWSC training in Annual Action Plan (AAP) 12-13 is 15,000 VWSC members.
- Proportion of NRDWP support funds spent on information, education and communication (IEC) and human resource development (HRD) activities.
  Status: ₹ 3,457.44 lac have been spent on IEC and HRD under NRDWP support fund out of 4,000 lacs i.e., 86 per cent are utilised.

The Way Forward

The Communication and Capacity Development Unit (CCDU) is conducting training of VWSCs to build their capacities to undertake the work of O&M of village water supply systems. Till today, CCDU has conducted the trainings of 63,500 VWSC members out of 1,37,655; 46 per cent target has been achieved so far. In 2012-13, capacity building of 15,000 VWSC members had been proposed. CCDU has joined hands with Indira Gandhi Panchayati Raj Sansthan (IGPRS), one of the pioneer institutes of Rajasthan, for capacity building of PRIs in the field of water and sanitation. Capacity building trainings have been started from state level to grass-roots level and PRI members are being trained in these trainings.

A state level Change Management Group (CMG) has been constituted consisting of 40 members. Rigorous training and exposure visits of this group are being conducted to prepare and equip them to pursue the new reform initiatives with respect to decentralised governance in water sector by CCDU, Rajasthan.

Similarly, 15 members of the CMG have also been created at divisional level to carry the agenda down the line at block and GP level. The state level group will train the divisional level CMG members. The emphasis in this training is to bring about change at personnel and community level with respect to new water reforms. Action plans have been prepared and they will be reviewed every quarter.

The government has issued direction to take up rain water harvesting and groundwater recharge on priority in every village water supply schemes. Rainwater harvesting tanks are being constructed in every school as well as at individual household and community level through Mahatma Gandhi National Rural Employment Guarantee Act (MGNREGA).

The Government of Rajasthan (GoR) has adopted water policy for the state. Integrated Water Resource Management (IWRM) has been adopted as major tool to manage conflict between different user groups. At present, IWRM plans are being formulated in 11 districts involving 2,798 GPs.

A communication campaign has been launched in the state to make people aware of the need for judicious use of water at household level as well as in agriculture sector. The existing capacity of PRIs in terms of village governance on water and sanitation facilities development and management in the world's largest democracy offers tremendous potential to make a positive difference. How this potential will be realised depends a great deal on how well they are empowered by the means of funds, functions and functionaries in terms of project implementation and monitoring process.

Many such projects implemented around the world have established that community-based participatory approaches involved robust informative, investigative and analytical monitoring tool, proved its importance in sustaining the project efforts and their wider use. There is a need for communities to be involved systematically in creating and managing their assets, analyse their problems and identify possible solutions to act, to marshal the resources necessary to implement and manage.

We need a robust monitoring and institutional mechanism and coordination among governments and water sector agencies to plan, develop and manage appropriate and sustainable drinking water supply systems. When a village level functionary reports the coverage of water and sanitation facility in a village or habitation, he should be made responsible for such reports and there should be cross and spot checks for such reporting systems. This should go on up to responsible officer at block, district and state level. The BRC will play a very big role in this matter in coming months. In Rajasthan, 249 BRC are being set up; 52 per cent recruitments have been completed. Similarly, efficiency of management information system (MIS) and computerisation at district level needs a thorough review in terms of their effectiveness and timeliness.

In order to address the sector challenges, mainly the slippages in habitations from 'fully covered' to 'partially' or 'not covered' status, the NRDWP guidelines emphasise the involvement of the PRIs and communities in planning, implementing and managing drinking water supply schemes, along with capacity building and M&E systems. Recently, the Ministry of Drinking Water Supply and Sanitation has prepared its long-term strategic plan (2011-2022) for ensuring drinking water security to all rural households. The strategic plan aims to cover 90 per cent of households with piped water and at least 80 per cent of households with tap connections during this period. The strategy emphasises achieving water security through decentralised governance with oversight and regulation, participatory planning and implementation of improved sources and schemes. Sustainable service delivery mechanisms are a central feature of the programme, with state institutions or ZPs implementing and managing large multi-village schemes, delivering bulk water to villages in water-stressed areas, and GPs implementing and managing in-village and intra-*panchayat* schemes.

The strategy highlights source sustainability measures, water quality safety, monitoring and surveillance, convergence of different development programmes, and building professional capacity at all levels. The main challenge now is the effective implementation and scaling-up of the proposed decentralised systems.

The state has not fared well in transfer of funds and functionaries to PRIs. The responsibilities have been transferred to some extent but cannot be discharged efficiently without funds and functionaries.

The state government is now contemplating the transfer of hand pump repair work as well as single village schemes to *panchayats* in two districts namely Udaipur and Alwar on a pilot basis. Since there are lots of administrative, financial and technical issues involved in the transfer of drinking water systems to panchayats, therefore, the state government has decided to test the transfer model first in selected districts and then upscale the same in remaining districts. The transfer will be wholesale, i.e., of funds, functions and functionaries. The model of transfer is being worked out in consultation with Panchayat Department, and it will be explained to all stakeholders in a workshop and the model will be explained to everyone. Based on the feedback of this workshop, milestones will be freezed for stages of transfer and subsequent replication efforts.

Conclusion

The GoR has already transferred five subjects to PRIs namely primary education, health, women and child development, social welfare and justice, and agriculture. The drinking water sector has still not been transferred to PRIs. The GoR principally agrees to transfer all the 29 subjects in the 11th schedule as per the 73rd amendment of the Constitution. The transfer of rural water supply to *panchayats* is proposed to be carried out in a phased manner to avoid any problem. The Janta Jal Yojna and water management within the village boundaries in all the major projects have already been handed over to *panchayats*. It is also proposed to transfer hand pumps schemes and single village schemes to GPs on pilot basis in Alwar and Udaipur districts with funds, functions and functionaries. A model is being worked out and it will be shared with all the stakeholders in a state level workshop very shortly. Based on the experience of this model, in these two districts, milestones will be fixed for transfer of other type of rural water supply schemes such as P&T, piped and regional schemes.

The state at present has scored only 17 points in MDI. The overall ranking of Rajasthan with reference to MDI in the country is 12th. There are certain indicators in the MDI which are required to be updated on the national website. This will result in further improvement in the MDI score. This indicates that the extent of decentralised governance in Rajasthan is better than what is reflected on the national website. The state government is committed to this concept but moving cautiously in view of the extreme importance of the sector because water is central to the entire developmental programming in Rajasthan which has traditionally been water starved.

References

PHED (n.d.). *Pragati Vivranika 2011-12*. Rajasthan: PHED.

Government of Rajasthan (2010). *State Water Policy, SWRPD*. February.

Websites:

www.ddws.gov.in

www.rajasthan.gov.in

7

JASWINDER KAUR and NAMIT KUMAR

Water Supply and Sanitation Systems in Punjab

A Study of Gram Panchayats in SAS Nagar District

Grass-roots level democracy is not a new concept, its origin dates back to the Vedic Age and has been fairly well documented in ancient scriptures such as *Rig Veda, Aitareya Brahman*, Kautilya's *Arthashastra* and writings of Buddhists' and Jain scholars. Under the British Rule, local government did not make much progress. However with the emergence of Mahatma Gandhi on the political arena, it slowly started regaining importance. Gandhi's advocacy of village democracy brought the issue to centre stage and laid the foundation for grass-roots level democracy in modern India. In 1992, a bill was passed as the 73rd Constitutional Amendment Act. The Act is a landmark legislation which brought empowerment to Panchayati Raj Institutions (PRIs) and ensured the transfer of power from the state to the PRIs to be exercised by the people. The Act added Part IX A to the Constitution. It is entitled as "The Panchayats" and consists of provisions from articles 243 to 243-O. In addition, the Act also added Eleventh Schedule to the Constitution.

The enactment of the 73rd Constitutional Amendment Act has highlighted the significance of grass-roots democratic decentralisation. As a framework of rural local self-governance, it became the tool to address the issues of both development and good governance at the grass-roots level. Further, as an instrument of decentralised governance, it focuses on enhancing the effectiveness and capacity building of local bodies. This change in the Indian political system was the result of a growing conviction that a big government cannot achieve growth and development without people's direct participation and initiative. This experience has brought with it the lesson that the success of the new Panchayati Raj system largely depends on the congruence of the perception and commitment of the people, their leaders and officials, about the role to be played by them in the new system. Contextually and theoretically too, the new Panchayati Raj system has been created as a model of self-governance.

Legislative framework under Article 243G provides 29 functional items to *panchayats*. Among these functional items, the provision of drinking water and sanitation draws importance as it has a serious impact on public health and it is a part of the global programme on poverty alleviation. The research paper discusses the delivery of water supply and sanitation services through decentralised institutions in Sahibzada Ajit Singh Nagar (SAS Nagar) District. It attempts to find out the problems faced by the *gram panchayats* (GPs) in performing these functions and is presented in three sections. Section I describes the condition of water supply and sanitation services in Punjab with special reference to two case studies in SAS Nagar district and Section II discusses the major findings of the study. Finally, Section III summarises the key issues and gives some suggestions followed by the conclusion.

Research methodology used both primary and secondary data for this study. The secondary data have been collected from various newspapers, academic journals, unpublished doctoral thesis and internet. For primary data, SAS Nagar has been selected as it is one of the fastest growing cities of Punjab and has proximity with the union territory of Chandigarh. The political leaders of the state are keen to make SAS Nagar similar to Chandigarh in all respects. The district includes 414 villages and three *tehsils*, namely, Mohali, Kharar and Dera Bassi. The Kharar *tehsil* is the largest administrative unit of the new district with 154 villages followed by Dera Bassi with 44 villages. Two GPs from Dera Bassi *tehsil* have been randomly selected. The study is based on a household level survey in two villages of the district including 100 respondents among which 8 were elected members and 92 were households. The data thus collected were critically analysed and presented.

I

Water and Sanitation in India

Water is the very basis of life and is the foundation for human survival and development. It is one of the most crucial elements in our national developmental planning for the 21st century. Sustainable and equitable use of water over millennia has been ensured by cultural adaptation to water availability through water conservation technologies, agricultural systems and cropping patterns adapted to different climatic zones, and conservation-based lifestyles. But in the last few decades, factors such as population growth, industrialisation and urbanisation, and the associated consumerist culture have interfered with the natural hydrological cycle of rainfall, soil moisture, groundwater, surface water and storage of all sizes. This has led to overuse, abuse and pollution of our vital water resources. Additionally, this has disturbed the natural cleansing capacity and quality of water.[1]

The provision of clean drinking water has been given priority in the Constitution of India, with Article 47 conferring the duty of providing clean drinking water and improving public health standards to the state.[2] The government has undertaken various programmes since independence to provide safe drinking water to the rural masses. Till the 11th Five Year Plan, an estimated total of ₹ 1,440 billion has been spent on providing safe drinking water. One would argue that the expenditure is huge but it is also true that despite such expenditure lack of safe and secure drinking water continues to be a major hurdle and a national economic burden. On one hand the pressures created by development are changing the distribution of water in the country, while on the other access to adequate water has been cited as the primary factor responsible for limiting development. The average availability of water is reducing steadily with the growing population, and it is estimated that by 2020, India will become a water stressed nation. Groundwater is

1. *http://planningcommission.gov.in/reports/genrep/bkpap2020/10_bg2020.pdf,* accessed on June 01, 2012.
2. *http://www.waterawards.in/suggested-reading/wateraid-drinking-water-quality.pdf,* accessed on May 22, 2012.

the major source of water in our country with 85 per cent of the population dependent on it.

The Census 2011 data have come in and people concerned with the water and sanitation situation of the country have found reasons to cheer because the figures point out a better coverage of drinking water facilities. As per the Census of India 2011[3] data, in so far as amenities available with the households, 43.5 per cent of households are using tap of which 32 per cent of water is treated and 11.6 per cent remains untreated. Forty-two per cent use hand pump/tube well, 11 per cent use wells out of which only 1.6 per cent of wells are covered while the rest uncovered. Lastly, 3 per cent use other source for drinking water. In rural areas, 51.9 per cent of the households depend upon hand pump/tube well, while 31.8 per cent have access to tap water. Further, as per Census 2011, 47 per cent of the households have latrine facility within premises out of which 36 per cent households have water closets and 9 per cent households have pit latrines.

From a simple analysis of the figures, it is revealed that though there has been an improvement in access to water (no doubt quality of water remains an issue), the progress with regard to sanitation has not been that encouraging. Toilets remain a distant dream still, as half of the households of the country still do not have access to toilets. Review of literature shows that there are variations in data, although the Census puts the people defecating in the open at 49.2 per cent, the Joint Monitoring Programme (JMP) of Millennium Development Goals (MDGs) assesses it at almost 61 per cent in March 2012. The MDGs target JMP for water and sanitation was released by the United Nation International Children Education Fund (UNICEF) and World Health Organization (WHO) reconfirming that sanitation is still a challenge and India was lacking behind by 11 years to meet the MDGs target by 2015 (Panda, 2012). Access to water, as per the Census claims, is much better. However, it is most important to note that a majority of the population of the country uses tap water which is untreated. Time and again emphasis has been laid on these issues. Of the 2.4

3. *http://censusindia.gov.in/, accessed* on April 30, 2012.

lakh GPs in the country, only 24,000—less than 10 per cent—are completely free of open defecation. Sixty per cent of the world's open defecation occurs in India. To change this situation, the Drinking Water and Sanitation Department has also seen its overall budget increase, with the Planning Commission approving a ₹ 36,000 crore allocation for the 12th Five Year Plan—up from ₹ 7,800 crore for the previous plan.[4] The 17th South Asian Association of Regional Cooperation (SAARC) Summit also puts forth that there is a need to formulate actionable framework to address the common challenges of sanitation and access to safe drinking water in Asia.[5]

Punjab's Water Supply and Sanitation Situation

Accessibility to water is not a problem in Punjab. Plenty of water, both groundwater and surface, is available. Most of the rural population depends upon untreated groundwater in order to meet their requirements of drinking water. Water is available in shallow as well as deep aquifers, but water from shallow aquifers is not safe for drinking due to contamination by industrial effluents and pesticides/insecticides from fields. In certain areas, groundwater contains high concentration of fluoride and chlorides. The growing population, increased requirement of water for irrigation purposes, particularly for rice crop, and demand of water for industrial use have lowered the groundwater table and as such residents have to resort to deeper wells/tube wells. The deeper tube wells are cost prohibitive for an individual.

The Habitation Survey 2003[6] revealed that 27.8 per cent of the rural households in Punjab were fully covered (by public water supply) in the sense that they have water supply level of 40 litres per capita per day. Another 37.9 per cent of the inhabitants were partially covered. Owing to the concentrated expansionary efforts by the state government under the 'Rajiv Gandhi National Drinking Water Mission,' the proportion of the fully covered habitats under public sector rural water supply was 93 per cent as on April 1,

4. *http://www.thehindu.com/todays-paper/tp-national/tp-newdelhi/article3507286.ece*, accessed on June 09, 2012.
5. *http://www.developindiagroup.com/SUMMITS.html*, accessed on June 04, 2012.
6. *www.global.ucsb.edu/punjab/journal_14_2/singh_singh.pdf*, accessed on May 25, 2012.

2006.[7] Thus only 7 per cent of the habitations in the state are yet to be brought under rural water supply mission. In contrast, as per the *Economic Survey 2012*,[8] Punjab has reported low (less than 50%) achievement in drinking water supply under Bharat Nirman. Hence, the researchers found a lot of variations in the data. In respect of sanitation facilities, 80 per cent of Punjab's rural population is below satisfaction level.[9]

In some areas, the main source of water for drinking is surface water which is made available from the existing network of irrigation canals. Although canal water supply has the potential to be used for drinking purposes, its supply is mainly utilised for irrigation of crops. The water supply is also subjected to breakdowns on account of canal closures for maintenance or other purposes. The water also needs proper treatment to eliminate suspended and dissolved solids carried through their catchments. The water supply for such areas needs to be augmented by other means and supply as such also requires proper treatment. Arranging and providing water supply has been the domain of residents/communities till 1950, when government initiated installation of tube wells to meet basic needs of drinking water of potable quality to reduce the impact of water borne diseases. Since then public service on provision of rural water supply has increased, but involvement of users and communities in planning and/or operation of the system has correspondingly decreased. GoI initiated various schemes for the fulfilment of the task in the form of Accelerated Rural Water Supply Programme (ARWPS) in 1972-73; 20-point Minimum Needs Programme (MNP) in 1975; Central Rural Sanitation Program; various types of matching grants; and providing loans under National Bank for Agriculture and Rural Development (NABARD) assistance. Planning and implementation of all these programmes have been through state institutions right from state level to *panchayat* level. The end result is that even though performance levels on providing and installation of sources and investment are

7. *http://www.global.ucsb.edu/punjab/journal_14_2/singh_singh.pdf*
8. *http://indiabudget.nic.in/es2011-12/echap-13.pdf*, accessed on May 25, 2012.
9. *http://pbdwss.gov.in/prwssp/Downloads/docs/Manual/Proc_Mnl/Procurement_Manual.pdf*, accessed on May 31, 2012.

recognisable, but being government dominated, the projects as a whole have become unsustainable over the operation period due to poor performance and absence of involvement by the beneficiaries thereby reducing the benefits of the investment.

Case Studies

SAS Nagar District had been carved out of areas falling in Ropar and Patiala District as the 18th District of Punjab on April 14, 2006. It is located in the north east part of Punjab and is a part of Patiala division. The district is spread over a geographical area of 1092.64 sq kms. According to the Provisional Population Totals of Census 2011,[10] the population of Punjab is 27,704,236 (2.77 crores); 62.51 per cent of the population is residing in rural areas and 37.49 per cent in urban areas. As per Population Census of India 2011, the total population of SAS Nagar is 9,86,147. The district has a sex ratio of 878, population density of 830 with a literacy rate of 84.9 per cent.[11] This section includes two case studies of villages, namely, Basma Colony and Basma in district SAS Nagar.

Case Study I (Village Name: Basma)

The village is situated on Punjab-Haryana border on the Banur-Ambala road. It has 400 households with a population of approximately 3,000. The household survey of the village revealed that 98 per cent of the population has water connection, but we know that access does not always translate into availability. The researcher observed that pipes and pumps might have been installed, but it does not mean that water actually flows out of the pipes for more than a couple of hours each day. Nor does it mean that the water that does come is safe or reliable, leaving infants and children, in particular, vulnerable to water-borne diseases. While talking to households, the researcher found that water supply of the village was disrupted since the past seven days due to a fault in the generator/motor at the water station. Such breakdowns in water

---

10. *http://censusindia.gov.in*, accessed on June 06, 2012.
11. *http://www.indiaonlinepages.com/population/punjab-population.html*, accessed on May 31, 2012.

supply were found to be quite frequent in the village. Further, there is only one water tank in the village which provides water to this village as well as to another nearby village, and 75 per cent of the villagers opined that this was one of the reasons for water shortage in their village.

Around 65 per cent of the villagers have hand pumps, while 35 per cent have submersible pumps in their houses. However, due to the rapid decline in the groundwater levels, hand pumps stop functioning in summers or people are able to draw just one or two buckets of water. The households who have submersible pumps in their premises do not face water shortages.

Moreover, it was also found that groundwater is being polluted due to the use of pesticides in the farms. Household survey also revealed that access to safe drinking water was an enormous challenge and the village *sarpanch* also corroborated these findings. She pointed out that while accessing drinking water continues to be a problem, assuring that it is safe is a challenge by itself. She observed that water-quality-related problems were being caused by pollution and over-exploitation. She also complained that she was unable to do anything as the top level officials at the district level did not give importance to the issues which were being raised by the GPs.

As far as sanitation is concerned, the *sarpanch* revealed that only 50 per cent of the households had toilets, while the rest of the households defecated in the open. Under the Punjab Water Supply and Sanitation Project, the *gram panchayats* have built toilets in around 30 per cent of the households. However, people were not used to these and did not feel like using toilets. They had converted these toilets into store rooms or had demolished them and used the bricks for some other household purpose. Around 50 per cent of the population opined that the toilets made in the houses were for women so that they do not have to go out while a majority of the male population defecated in the open.

Findings also revealed that the people were unaware about the ill-effects of defecating in open while one-third of villagers were yet to be connected to safe drinking water services and that half the rural households needed proper and hygienic sanitation facilities.

A visit to the villages shows improvement in other infrastructure like streets and cremation ground maintenance. Further, it can be said that since decentralised institutions have performed well in managing other infrastructure and services in the village, therefore, if given a chance, they can also succeed in managing and improving water and sanitation facilities.

Case Study II (Village Name: Basma Colony)

This village is situated near Basma village and has around 50 households making up a population of about 300 residents. The survey revealed encouraging situation as almost all the households had access to toilet in their houses. But effluent from septic tanks could be found flowing in open drains. This, as everybody knows, not only poses serious health hazards but has also could lead to degradation of the environment. Access to drinking water was found to be good, as all the households were connected with pipelines for water supply. But as discussed earlier that having water connections does not necessarily mean that households actually get water supply. Infact, a vast majority of the households (90%) revealed that though pipelines were in place, water supply was negligible in the summers due to recurrent faults in the water station. In case of breakdowns in water supply, people are using groundwater, and although every house had its own hand pump, only 10 per cent of the households have submersible pumps in their houses. Unfortunately, due to the open sewerage system and seepage of this waste is leading to pollution of undergroundwater resulting in health hazards for the village community.

According to the village *sarpanch*, regular water supply was a distant dream for the villagers. The *sarpanch* informed that the Fatehagrh Sahib District had taken a step in the direction of empowerment of *panchayats* by giving the authority to *panchayats* of two villages (Chunni Khurd and Chunni Kalan) to operate and maintain the rural drinking water supply system. The *sarpanch* suggested that this could be done in all the districts.

A majority of people believe that sanitation programmes and projects have not yielded the desired results owing to lack of involvement and commitment from both communities and external agencies—government or non-government. Besides there were also lapses in technology, planning, implementation, supervision, support and, above all, accountability. They suggested that in order to make services sustainable and programmes successful, the quality of construction work should be improved, corruption should be controlled, and an effective operation and maintenance (O&M) system should be established.

A common but predominant message emerged from the household surveys across the belt was that political leadership must take a collective resolve in the region to promote right to sanitation and assure dignified lives to people by providing them and their children with a disease-free and healthy environment.

II

Major Findings

The primary and secondary data revealed that the *panchayats* were in poor financial condition. Consequently, they were not in a position to afford huge investment in infrastructure facilities like water supply, sewerage, drainage, solid waste management and power supply. Punjab Water Supply and Sewerage Project came as a significant move to improve access to drinking water and sanitation situation; however, comprehensive planning and effective monitoring were essential to take this scheme successfully to its logical conclusion. Substantial capacity building and strengthening will be a pre-requisite in empowering the *panchayats* with the necessary skills to take the village into the future. The researchers found that although the *sarpanchs* were aware of the exact situation of their village, they were unable to do anything as they had little autonomy. However, experience shows that wherever elected representatives had been given power, they had performed well.

But water scarcity was not prevalent everywhere and that some villages from the same district were also enjoying round-the-clock water supply (e.g., Singhpura and Sitabgarh villages).[12] Further, there was huge gap between demand and supply of services, and the GP was unable to cope with the pressure. Although *panchayats* have been assigned several powers and functions under the 73rd Constitutional Amendment Act, 1992, the study of GPs shows that in practice they have little autonomy.

When asked to give their personal opinion about water and sanitation situation, the *sarpanchs* of both the villages told that the situation was poor. They said that people in villages wanted a 'clean' and 'healthy' environment for themselves and their families. They aspired for dignity, privacy and freedom from a life of shame and embarrassment. They wanted functional toilets, waste water disposal systems, and adequate and regular arrangements for disposal of solid waste. Further, the *sarpanchs* opined that although government publications highlighted that the Punjab government was spending a lot of funds to improve this situation, very little was actually being done at the ground level.

Recently, the GoI has increased the spending on toilets for rural areas, increasing the amount to be spent for constructing a household latrine from the existing ₹ 4,600 to ₹ 10,000. Now, GPs will receive anywhere between ₹ 7 lakh and ₹ 20 lakh for this purpose. The government has also mandated that anyone receiving funds to build a house under the Indira Awaas Yojana must construct a toilet as well (*The Hindu*, 2012).

III

Key Issues and Suggestions

The scarcity of drinking water and lack of toilets is a huge issue that has to be addressed in the country. With water getting scarcer, this imbalance will further grow. If the country wants sanitation coverage to catch up with that of water, these two have to be linked and not separated. The country is starting to recognise that water and sanitation are basic human rights, as has been proposed in Draft National Water Policy 2012. The key issues emerging from the

12. *http://www.tribuneindia.com/2011/20111216/punjab.htm#10,* accessed on May 31, 2012.

household surveys in different village *panchayats* has been discussed as follows.

- *Groundwater Extraction:* Punjab faces a serious problem of over extraction of groundwater, primarily due to the widespread use of irrigation tube wells coupled with non-supply of drinking water. The biggest problem that the state is going to face in the future is lack of adequate groundwater. There is a pressing need for the state government to enact and implement an act for controlling groundwater extraction in the state. Rain water harvesting schemes should also be launched in rural areas.
- *Time-bound Implementation Plan*: Punjab has made limited progress in setting up laboratories for testing water quality. This is an area for concern given the widespread problems of water quality in the state. Thus, there is a need for a specific time-bound implementation plan for improving the physical infrastructure for water quality testing as well as for regular periodic checks of all sources.
- *Autonomy to PRIs*: The lack of capacity and autonomy of PRIs resulted in the failure of many projects. Measures envisaged under the 73rd Constitutional Amendment Act should be adopted to strengthen the third tier of the government to provide them full autonomy. Transferring O&M functions without appropriate strengthening of the PRIs is likely to lead to less than desirable outcomes. In other words, intensive capacity building measures that include improving skills, attitudes and knowledge of PRI staff would have to precede successful implementation of different schemes in the state which is the main objective of decentralised governance.
- *Coordination and Accountability Mechanisms:* There is an urgent need for coordination between the institutional structure at centre with that at the state and district level. Programmes are implemented through multiple agencies and issues related to coordination, communication and accountability are prevalent. To overcome these

issues, proper mechanisms should be put in place by the government so that each agency is held accountable for its own work.

- *Documentation*: Best practices should be documented and priority must be given to adopting same state best practices. This will help others to adopt the same strategy to achieve best results.
- To identify minimal conditions of governance that are necessary to allow political and economic development to occur (Andrews, 2012).
- *Non-Willingness of Officials*: Lack of public sector investment and negligible or no political commitment at higher level, which has otherwise brought a change in the developed world, is a major stumbling block to upgrade the rural areas. Hence, local officials should be encouraged to take active part in implementing such schemes. They should become active agents of social change.
- *Social Audit*: The lack of regular review mechanisms frees implementing agencies from accountability. Thus, regular review of implementation should be conducted by the state government. The involvement of the citizen, as is the case of 'Social Audit' in MGNREGAs, is highly desirable to deepen the degree of accountability.
- *Addressing Local Issues*: The local issues raised by the GP were given least importance by the government officials. Hence the main motive/objective of decentralised governance has been kept somewhere behind the curtain. Thus, importance should be given to local issues.
- *Peoples' Participation*: There is a lack of people's participation at the grass-roots level in the implementation of such schemes owing to reasons such as illiteracy, caste rigidities, etc., which compromises the very purpose of decentralised governance. Hence, community involvement is the first and foremost prerequisite in the creation and maintenance of such facilities.

Conclusion

This research paper includes case studies of two GPs in the district to review the water supply and sanitation services. An attempt has been made to find out the perceptions of the villagers and the village *panchayats*. The research paper revealed that Punjab is a water surplus state, however quality of potable water is an issue. Access to drinking water was found to be good as all the households were connected with pipelines for water supply. It is found that though pipelines were in place, water supply was negligible in the summers due to recurrent faults in the water station. In that case, people are using groundwater (hand pump and submersible pumps). The survey revealed encouraging situation as almost all the households had access to toilet in their houses. But effluent from septic tanks could be found flowing in open drains. Unfortunately, the open sewerage system and seepage of this waste is leading to pollution of undergroundwater resulting in health hazards for the village community. The researcher opines that if the country wants to improve further more in this, then they should not treat both the subject as separately, as both affect each other. In quintessence, it can be concluded that decentralised governance as a model has become base for the development of the rural areas and their citizens as well. Emphasis should be given to maximising innovation, and the most important aspect is to strengthen the innovative forces and ideas which lead to development.

References

Andrews, Matt (2012). "The Logical Limits of the Best Practice Administrative Solutions in the Developing Countries", *Public Administration and Development* 3(2), May.

Panda, Ranjan K. (2012). "Water-Sanitation Nexus More Complex than Toilet—Cell Phone Comparison", *Yojana*: 21-24, May.

The Hindu (2012). "Now More Funds for Toilets in Rural Areas", *The Hindu*, June 09.

8

KAPIL LALL

Rural Water Supply and Sanitation Sector

Initiatives in Uttarkhand

Introduction

Massive investments have been made in the rural water supply (RWS) and sanitation sector during the last two decades. However, providing adequate and potable drinking water on a sustainable basis to all persons living in the rural areas is a major challenge being faced by water supply managers and policy planners of our country. The sustainable delivery of RWS and sanitation service has been a cause of concern for policy planners of the country. This has led to a paradigm shift in the way the government now approaches the RWS and sanitation sector. The Government of Uttarakhand (GoUK) had been equally concerned to ensure sustainable development in the delivery of rural water supply and sanitation (RWSS) service.

In 1999, India embarked on a significant reform of the RWSS sector. It sought to replace the top-down approach of the past with a decentralised approach that placed the people in the driver's seat of development. This implies that communities should choose the service they want and are willing to pay for, and be responsible for the planning, construction and maintenance of their own systems. In keeping with this policy, the government's roles have changed from being a 'service provider' to that of a 'facilitator.'

The policy planners of the GoUK diagnosed the unsustainable delivery of RWSS issue in the context of the state. The findings revealed that the planning and implementation process in the delivery of RWSS remains engineering oriented and does not involve communities in basic decisions regarding type and level of service to be provided. Professionals in the sector, though belated, realised that the delivery of RWSS is not just a technical issue, but has social community, gender, financial institutional, political management and environmental aspects as well.

The State of Uttarakhand

The topography of Uttarakhand is characterised by hilly terrain with altitudes varying between 100 meters and 7,800 meters above mean sea level. Correspondingly, the settlement pattern, population spread, and water supply sources, and consequently the degree of difficulty vary substantially. The total area of the state is 53,480 sq km of which 87 per cent territory is hills and 13 per cent are plains. The average population density per sq km of the state is 159; however, it ranges between 41 (Uttarkashi) to 817 (Haridwar).

The Uttarakhand state, which bifurcated from Uttar Pradesh in November 2000, has a population of 8.48 million as per the 2001 census, of which 6.4 million (75%) live in the rural areas, spread over the 7,562 *gram panchayats* (GPs) in 16,623 villages and 39,180 habitations. As on April 1, 2011, out of total 39,180 habitations, 75 per cent (29,352 habitations) are categorised as fully covered (FC) with functioning water supply schemes, 19 per cent (7,609 habitations) as partially covered (PC) and about 6 per cent (2,219 habitations) as not covered (NC).

There are three agencies engaged in the delivery of rural water supply and sanitation service (i) Uttarakhand Peyjal Nigam (UJN); (ii) Uttarakhand Jal Sansthan (UJS) and (iii) Project Management Unit (PMU) Swajal. UJN is the main implementing agency for construction of water supply schemes under National Rural Drinking Water Programme (NRDWP) and Uttarakhand Rural Water Supply and Sanitation Project (URWSSP). UJS mainly operates and maintains single village (SVS) and multi-village schemes (MVS) which have not been devolved to Panchayati Raj Institutions (PRIs) for operation and maintenance (O&M). PMU Swajal is a society registered under the Societies Registration Act primarily for implementation of URWSSP and GoI financed

Total Sanitation Campaign (TSC) and Water Supply and Support Organisation (WSSO) under NRDWP.

The degree of difficulty in accessing RWSS services are extremely high in the hilly regions of the state. The limiting factors for timely completion of infrastructure works include: (i) short working season due to frequent landslides; (ii) inhospitable terrain; (iii) remoteness of gravity/surface pumping water sources from road head; (iv) delay in forest clearances (65% forest cover); (v) temperature below freezing point in high mountain areas during winter; and (vi) weak financial and technical capacity of contractors for execution of turn key works.

The GoUK is committed to implement the sector reforms initiated by GoI and empowering the PRIs as envisaged in the 73rd Constitutional Amendment Act. It has prioritised RWSS as a key area of its development agenda.

The "GoUK RWSS Vision" envisages universal coverage of safe and potable water and sanitation. The vision also delineates the renewed institutional structure of the state as: "The rural local government in partnership with rural communities; shall plan, design, construct, operate and maintain their water supply and sanitation schemes; so that they get potable water and accrue health and hygiene benefits. The government and its sector institutions together shall act as a supporter, facilitator and co-financier, technical assistance and training provider and also support large scale construction works and sectoral contingencies."

The Uttarakhand Rural Water Supply and Sanitation Project (URWSSP)

This project URWSSP is the first of its kind across the globe in the sense this is the first project in the RWSS sector which is based on sector-wide approach (SWAp). The project covers entire rural areas of the Uttarakhand state. The project is facilitated by the Department of Drinking Water, GoUK, and executed by three agencies namely UJN, UJS and PMU Swajal. The project has a total budget of $ 224 million of which the International Development Association (IDA) credit is limited to a maximum of $ 120 million.

The project period is November 2006 to June, 2014. The project aims to benefit about 1.2 million people or about 20 per cent of the rural population and improve sanitation in about 30 per cent of rural communities to be declared free of open defecation.

Development Objectives

The development objective of the project is to improve effectiveness of RWSS services through decentralisation and increased role of PRIs and involvement of local communities as a SWAp in Uttarakhand. Key principles underpinning the project strategy are: (i) voice and choice of the people—informed choice; (ii) helping people to help themselves; (iii) self-selection and demand driven—local communities select themselves, to participate or not in the project; and (iv) decentralised decision making—project is fully aligned with the three-tier PRIs which is in conformity to the 73rd Constitutional Amendment.

The project is strictly premised on the principle of 'subsidiarity' *viz.*, that a particular decision/activity will be made/take place at the lowest most appropriate level. Accordingly, the habitation is the focal point for planning, implementation, O&M of the water supply scheme, rather than revenue village or the GP. User water and sanitation sub-committees (UWSSCs) is, in most cases, the primary management units exercising control over resources and decision making. The project makes extensive use of existing government systems for funds flow, financial control and financial reporting.

State Level Institutional Arrangements

The key institutions that play a significant role in implementing URWSSP are at the state, district and village levels. Department of Drinking Water (DDW) of the state government is the nodal agency for the sector and coordinates with the sector stakeholders and other sector ministries and departments such as Health, Eduction, PRIs and Watershed Management. State Water and Sanitation Mission (SWSM). SWSM is the highest policy-making body in the sector. Its functions include: (i) oversee, disseminate and monitor the implementation of policy decisions undertaken by SWSM; (ii) compile, review, and approve the annual plan and budget allocation for the sector for monitoring; and (iii) appraise and approve high-cost MVSs that are beyond the prescribed limit of District Water and Sanitation Mission (DWSM). The functions of the PMU include: (i) coordinate and implement the new capital investments for SVSs and (ii) undertake informations, education and communication (IEC) campaigns and the capacity-building activities. The functions of UJN and UJS, along with their district agencies, include: (i) plan, construct (other than for intra-village investments), operate, and maintain the MVS according to the Memorandum of Understanding (MoU) and (ii) provide technical support to the PRIs and the UWSSCs for SVS, if desired by the latter.

District Level Institutional Arrangements

- DWSM headed by the Chairman *zilla panchayat* (ZP) has the mandate for (i) selection of habitations, approval of annual budget, review of programme and dispute resolution and (ii) appraisal and approval of proposals up to ₹ 50.00 lakh and recommendations exceeding ₹ 50.00 lakh to SWSM and GoUK.
- *District Water and Sanitation Committee*: DWSC is headed by Chief Development Officer and other district level officers of related departments and its functions include: (i) selection of support organisations for community mobilisation; (ii) Review and approval of schemes proposed by GP and UWSSC; and (iii) approval of district schedule of rates for engineering items/works/materials.

Village Level Institutional Arrangements

- *Gram Panchayat* : The GP is the primary management unit exercising control over resources and decision making. The GP mobilises and supports the formation of UWSSC to ensure participatory approach. GP empowers and provides capacity support to the UWSSC; ensures O&M and cost recovery of the scheme; fund flows, scheme approval, accounts management, auditing, monitoring and evaluation (M&E), and conflict resolution.

- UWSSC consists of beneficiaries of the scheme is responsible for implementing all the phases of the project cycle including procurement of materials, tariff fixation and revision, community contributions, accounts management and O&M of SVS and intra village MVS. Panchayat Raj Act of the state bestows legal right to the UWSSC, as per the 73rd Constitutional Amendment.

Fund Flow Mechanisms

The fund flow mechanism in the project is community friendly. The SWSM, supported by PMU, is responsible for overall financial management, funds flows, financial reporting and submission of reimbursement claim to the World Bank. The funds from the state government flow directly to the state level implementing agencies (PMU, UJN and UJS). The funds from the state level are transferred to the district level institutions of PMU, UJN and UJS, based on Annual Work Plans and Budgets. For the construction of common infrastructure required for complex MVS, the funds flow to the UJN. Funds from district level institutions of all the three agencies are transferred in three tranches to the Gram Nidhi Account of GP for construction of SVSs and intra village works of MVS. GP receives the funds in Gram Nidhi (GP account) from the district implementing agency (DIA) and transfers the amount by cheque to the capital fund account of UWSSC. The release is linked to the implementation progress review based on physical and financial progress report submitted by the UWSSC and GP in case of SVS and by the UJN in case of MVS. Separate capital and O&M bank accounts are opened and operated by each UWSSC. UWSSC collects initial user contribution towards capital cost and also collect retain user charges to pay for ongoing maintenance.

The Project Cycle for Single Village Schemes (SVSs)

SVS includes four phases: pre-planning (3 months), planning (6 months), implementation (6-18 months) and O&M (4 months). The duration of each phase depends on the scheme size, technology type and the time it takes to mobilise the communities.

- *Pre-planning:* The outputs of pre-planning phase include selection and approval of needy GPs and support organisations (SOs). The selection of GPs and SOs ultimately culminate into signing of the planning phase agreement.
- *Planning:* During planning phase, the SO interacts with the village community members and mobilises them towards the roles and responsibilities of the village communities, the User Water and Sanitation Sub-committees (UWSSCs) and the GPs for effective community-driven planning of the water supply and sanitation schemes—an assessment of community needs and plan of action.
- *Formation of UWSSCs:* UWSSCs are formed as per the government notification for planning which plays a lead role in the whole planning process. The SO facilitates the communities in choosing their representatives to form UWSSCs. To provide a statutory status, the UWSSC is formed as a sub-committee of the GP. The SO also helps the UWSSC to draft the bye-laws for functioning of UWSSC in O&M phase. Apart from this, the SO assists in opening two bank accounts—one for capital fund account and another for the collection of upfront cash contribution for capital cost and O&M cost.
- *Collecting Cash Contributions:* One of the requirements of progressing from the planning phase to the implementation phase is that the community should collect upfront cash contribution as per the cost sharing rules. The UWSSCs shall collect upfront capital cash contribution from their respective user group. The UWSSC maintains a register for the purpose. Confirmation of the requisite collection of upfront capital cost has to be communicated to the DIA. In addition, 50 per cent of estimated one year's O&M charges are also to be collected upfront and deposited in each

UWSSCs O&M A/c and a confirmation for this has to be sent to the DIA.

- *Preparation of the Implementation Phase Proposal*: The Implementation Phase Proposal has two plans: (i) Technical Plan and (ii) Community Empowerment Plan. These plans are conceived and prepared after detailed discussions with the GP, User Group and entire community level. The Technical Plan (in form of Detailed Project Report) has three sub-plans (i) Water Scheme Layout Plan; (ii) Village Latrine Plan; and (iii) Catchment Area Conservation & Management Plan. This plan includes the technical design, bill of quantities and the financial proposal for construction activities. The Community Empowerment Plan (CAP) includes Hygiene & Environmental Sanitation Awareness (HESA) and Women's Development Initiatives (WDI). These plans clearly spell out the activities to be undertaken together with the expected time. The plans are drawn by the community as per their demand and needs are village-specific.
- *Reviews by District Water and Sanitation Committee (DWSC):* All the DPRs are scrutinised at this level to ensure that all the project conditions have been fulfilled. After scrutiny, the DPRs are put up before the District Water and Sanitation Mission for approval up to ₹ 50 lakh and for recommendation in case of DPR cost exceeding ₹ 50 lakh to the SWSM.
- *Signing the Implemenation Phase Quadruple Agreement (IPQA)*: After approval of the SVS, IPQA is signed among DIA, SO, GP and UWSSC for implementation of the scheme. In case of MVS, construction of common assets of MVS head work, distribution main up till the head of each covered village by the sector institution and construction of intra village works are done by the GP/UWSSC.
- *Implementation:* The duration of the implementation phase ranges between 6 months to 30 months depending upon the scheme size, technology type and the time it takes to

mobilise the community. The procedure is as follows: The GP receives the funds in Gram Nidhi from the DIA and manages the funds received for construction of drinking water scheme, sanitation and catchment area conservation works and transfers the amount received in Gram Nidhi by cheque to the capital fund account of UWSSCs within 15 days. Separate ledgers are maintained by GP for different UWSSCs. The Gram Nidhi account is operated by *gram pradhan* and Secretary of GP. The maintenance of accounts of the funds received for drinking water and sanitation scheme is done at the level of GP according to the proforma/formats prescribed by the accountant general. The GP ensures auditing of Gram Nidhi account and the UWSSCs ensure auditing of the capital fund account and O&M account of the UWSSCs. The UWSSC (with the help of the GP and SO) is responsible for the overall implementation of construction and community development activities. The construction activities are to be implemented in accordance with the approved DPR. The UWSSC (with the help of GP) is responsible for managing the procurement of materials and their storage in case of SVS and intra-village works of MVS. The DIA provides training to the SO staff to enable them to train communities. The SO is responsible for training and capacity building of GP/UWSSC members.The DIA is also responsible for appointing third party construction supervision agency and an auditor to conduct an independent audit of records pertaining to the agreement. Balance 50 per cent community contributions are also collected during this phase. After completion of the works, Implementation Phase Completion Report is prepared and signed by all the parties to the agreement.

- *Operation and Maintenance:* The duration of the O&M Phase is 4 months. The O&M phase includes the following activities. The DIA provides technical assistance to the UWSSCs after commissioning of the water supply schemes to place the O&M system in order. Trainings at the GP/UWSSC level are conducted by DIA directly or by outsourcing. The O&M system comprises the technical, financial and the institutional systems. After establishing O&M system and completing all the activities stipulated in the agreement, the DIA formally exits from the GP.

The Project Cycle for Multi-village Schemes

Each scheme cycle includes four phases, including pre-planning (2 months), planning (6 months), implementation (12-30 months) and O&M phases (4 months). The duration of each phase would depend on the scheme size, technology type and the time it takes to mobilise the communities. The pre-planning and planning phase activities are the same even for MVS. However, implementation phase is different which includes: (i) construction of the common assets of MVS head work, distribution main up till the head of each covered village by the sector institution and (ii) construction of intra-village water supply schemes, environmental sanitation works catchment area protection works by GP/UWSSC through community engineer. The responsibility for O&M of common assets rests with UJN and with UWSSC for intra-village works.

Innovations at a Glance

It is common knowledge that sector-wide reform cannot be ushered in overnight. However, the steps taken must head towards the sector agenda and goals. Following are the innovations and initiatives that the government has taken to steer the reform process in the state:

- *Habitation as the Unit of Planning:* The habitation is the focal point for planning, implementation and O&M of the water supply scheme, rather than revenue village or the GP.
- *Legality to the UWSSC*: Appropriate provisions have been made in the Panchayat Raj Act of the state to bestow legality to the UWSSC as per 73rd Constitutional Amendment.
- *Model Bye-laws for the Functioning of UWSSC*: Model bye-laws have been put in place for smooth conduct of business of UWSSC.

- *Fine Tuning Support Organisations' Selection Criteria:* SOs are selected using comprehensive eligibility criteria in order to weed out 'Bad quality' from 'Good quality' SOs.
- *Integrated Delivery of Water Supply Services:* The water supply schemes are integrated with catchment area programmes, household and village environmental sanitation programmes, solid waste management, and health and hygiene awareness promotion programs, in order to maximise water supply and health benefits to the communities.
- *Constitution of District Level Technical Review Committee (TRC):* A district level TRC is constituted to review the technological options proposed by the various User Groups, analyse the justification of SVS/MVS, and to finalise the scheme identification plan.
- *Single Set of Schedule of Rates:* The district schedule of rates for various engineering items/works/materials (local and non-local) are jointly prepared by all the three implementing agencies and approved by the District Water and Sanitation Mission (DWSM).
- *Capacity Building of Stakeholders:* Special training programmes have been developed and implemented through the local training institutions to sensitise the state, district and village stakeholders and functionaries regarding the RWSS SWAp programme.
- *Information, Education and Communications (IEC):* IEC activities through workshops, cross-visits, competitions, IEC stalls, TV programmes, etc. A couple of short documentary films have been prepared on the good practices currently being implemented under the project. A quarterly newsletter 'Swajal Samachar' is regularly published for dissemination of project philosophies and principles as well as experience sharing and lessons learnt.
- *Insurance of the Schemes*: Insurance is an instrument through which sustainability of the water supply scheme is

ensured during the construction phase in natural calamities dominated state like Uttarakhand.

- *Social Audit Committee:* At the GP level, a Social Audit Committee has been constituted for rectifying the problems related to construction of scheme. The main role and responsibilities of Social Audit Committee is: (i) to ensure that all the committees follow Procurement Manual; (ii) report any violation or deviation of rules to GP; (iii) monitor the adherence of project principles and rules in the selection of beneficiaries; and (iv) oversee the implementation of sub-projects and all decisions of UWSSC.
- *Monitoring and Performance Tracking*: A web-based robust monitoring and performance tracking system is in place to monitor processes, inputs, outputs and outcomes, including indicators of change under the project. The URL is *http://swajalsis.uk.gov.in/;* password is not required for access of online data reports. The reports, *inter-alia*, which can be seen include: (i) pre-planning phase, (ii) planning phase, (iii) DPR detail, (iv) implementation phase, (v) contract details, (vi) baseline data, (vii) MVS and (viii) O&M
- *Elaborate Exit Strategy:* The exit or *Atmarpan* is a process of gradual withdrawal of the SO and the DIA from the village in anticipation that the community has been empowered to operate and maintain its water supply and sanitation schemes on its own. The project has put in place an elaborate exit policy to ensure sustainability of the created assets.

Sustainability Evaluation Mechanism

The project has designed a Sustainability Evaluation Exercise (SEE) to assess the sustainability of completed schemes which are more than one year old.

- Independent reviews are an integral part of the project processes including concurrent monitoring during the scheme planning phase, third party construction quality checks, technical audits, social audits and grievance redressal during the scheme implementation phase.

- *Grievance Redressal Mechanism:* An effective system to redress grievances has been established at the village and district levels. Any issue raised by any member of the local community is first discussed and deliberated at the UWSSC/GP level, where most disputes/issues are resolved generally by open meetings. All the grievances related to water source use, location of tap stand and type of technology establishing water points are being sorted out in community-wide meetings. All the grievances related to material procurement, quality of works, workmanship and O&M phase are being sorted out in UWSSC, Social Audit Committee and community-wide meeting. The grievances which are not resolved at the community level are referred to DWSM/SWSM level for redressal.

Post-Implementation Support

As per National Policy, state vision and SWAp approach, the PRIs have to manage, operate and maintain the rural water supply services in villages and habitations of Uttarakhand. The state government has designated UJS as the nodal agency for providing back-stop support to all commissioned rural water supply schemes implemented under supply- or demand-driven approach. Apart from this, the UWSSC members, operators and communities are regularly trained by the Communication and Capacity Development Unit (CCDU) of GoI. Needless to mention, the District Project Management Unit (DPMUs)/district divisions of UJN and UJS provide technical assistance to the UWSSCs as and when required by them, especially in the case of emergencies.

Impacts

By February 2012, over 4,300 habitations have completed their own water supply systems under the SWAp programme, benefiting some half a million rural residents. Following are the major impacts:

- *Building Local Capacity*: The new decentralised approach has been received with a high level of enthusiasm from local governance institutions. By building the capacity of these institutions to design, implement and manage their own works—in addition to managing procurement, finances and accounting—they have been equipped to take on other development projects in their areas, whether under MGNREGA or any other programme.
- *Eliminating Layers of Bureaucracy:* Decentralisation to local communities has eliminated layers of bureaucracy, reducing the scope for leakage of funds—an estimated 40 per cent of funds deployed, time and expenses involved in lodging complaints with the authorities, paying bills, getting corrections in wrong bills, and getting sanctions for new water connections. Repairs are now much quicker and more long-lasting because of the strict supervision exercised by local communities.
- *Savings for the State Exchequer:* Detailed assessments show that by decentralising service delivery to local communities, the state government's savings include: savings in institutional costs and O&M costs. Thus the state government has to provide lesser budgetary support through State Exchequer as GP meets the O&M expenses including energy costs.
- *Greater Availability of Time for Line Departments:* GP/UWSSC takes up the full responsibility of O&M, with back-up support by the UJS, as and when required. Thus line departments have more availability of time for managing complex schemes and lesser management problems.

Advantages of Communities

- Lesser dependence on outsiders for water supply service i.e., Gram Swaraj (village self-rule)—leading to 'community pride' and 'community control'.

- Reduction in coping cost as well as more availability of household space due to elimination of storing water in larger containers
- elimination of travel time and travel cost in lodging no-water complaints, making payment of bills and getting sanction of new water connection
- Improvement in mean time between failure (MTBF) due to better quality of repairs and strict control/supervision by the community.
- Reduction in women's drudgery of about 2-3 hours per day leading to better quality of life i.e., utilisation of their quality time in caring their children and leisure time besides income generation in some cases.

Conclusion

The URWSS project has demonstrated the success of demand-driven community participatory approach by forming and empowering the communities to plan, construct, operate and maintained their own rural water supply schemes. The state has transferred the control of 'funds' functions and functionaries' down to the lowest rung of the local governance infrastructure—GPs transforming the way rural water supply infrastructure is created and managed in the state. Water supply managers and policy planners, though belated, have realised that the delivery of RWSS is not just a technical issue, but has social, community, gender, financial, institutional, political, management and environmental aspects as well. The state has benefited from initiatives involving communities in developing and managing their rural water supply and sanitation schemes. In short, the state government's role has changed from the a 'service provider' to that of a 'facilitator.'

9

V.C. GOYAL and R.D. SINGH

Source Sustainability in Drinking Water Scheme

The Role of Hydrology

Introduction

Water supply in India, especially in rural areas, is a challenging situation. Sector institutions have been effective in building infrastructure and most habitations in rural India are now covered with access to hand pumps or piped water supply systems. However, there are also high rates of slippage of hand pumps and piped schemes due to lack of ownership, poor O&M, sources going dry or lowering groundwater tables, sources becoming quality affected and increasing populations.

Besides availability, access to drinking water is a continuing serious water issue, particularly for the poor. Access to water depends on a complex of factors, including the local availability of water, its quality and the economic ability to obtain it. Some 1.6 billion people worldwide have limited access to water, even though it is locally available. For example, many poor communities in deserts may need to travel long distances to get surface water because they cannot afford to drill wells to get the groundwater flowing beneath their feet. Even in water-rich areas, if the local water is polluted and there are no economically viable alternatives, access is effectively denied. Over one billion people in the world lack access to safe water supplies.

Water supply systems require a variety of resources in order to be designed and managed sustainably. These can be grouped into five categories (Reed, 2001):

- human (education, skills),
- financial (credit, banking services),
- physical (pipes, pumps, dams, spare parts),
- natural (water, land, building materials), and
- social (community institutions and organisations).

The revised National Rural Drinking Water Programme (NRDWP, 2010) of the Government of India (GoI) aims at "providing every rural person with adequate safe water for drinking, cooking and other domestic basic needs on a sustainable basis. This basic requirement should meet minimum water quality standards and be readily and conveniently accessible at all times and in all situations." The major thrust of the rural water supply programmes is to ensure provision of adequate drinking water supply to the rural communities. Hitherto the role of the government was that of a service provider, with minimum involvement of the community in the planning process and implementation activities. With the increase in demand due to public aspirations and decrease of availability of safe water, it was considered essential to involve the community in the programme.

Water quality management, especially in rural context, can be meaningfully achieved through community-based approaches. This necessitates integration of water quality initiative with sanitation and hygiene and ultimately to the health of the people. The economic value associated with benefits of an intervention will motivate the community to accept (and adopt) the intervention (James, 2003). Community-led water supply projects need to be managed as locally as is reasonable so they are based on local demands and are sustainable. This requires sharing of resources and skills, including knowledge on the hydrology principles which is vital for long-term sustainability of the water availability (both in quantity and quality).

Sustainability in Drinking Water Schemes

Sustainability of drinking water sources and schemes is a process which facilitates the existing/new drinking water supply projects to provide safe drinking water in adequate quantity, even during distress periods, duly addressing equity, gender, vulnerability, convenience and consumer preference issues, through conjunctive use of groundwater, surface water and roof-water harvesting. The main aim of providing sustainability in drinking water schemes is that such schemes will not slip back from universal access of safe drinking water to the community throughout the design period of schemes. The ultimate aim in water supply schemes is to achieve household level drinking water security. Adoption of appropriate technology, revival of traditional systems, conjunctive use of surface and groundwater, conservation, rain water harvesting and recharging of drinking water sources, are all considered integral components for achieving drinking water security.

Broadly, the four elements of sustainability are: (i) source sustainability: ensuring availability of potable drinking water in adequate quantity throughout the year; (ii) system sustainability: optimising the cost of production of water, devising proper protocol for O&M, capacity building of PRIs and awareness generation; (iii) financial sustainability: proper utilisation of funds; and (iv) social and environmental sustainability: proper project management and involvement of all key stakeholders.

The GoI in 1999 initiated institutionalising community participation in the implementation of rural drinking water supply schemes for ensuring sustainability of the systems. The approach relied on providing the cost sharing for O&M responsibility by the end users. Later on it was realised that sustainability in the drinking water schemes encompasses many more components (e.g., protecting sources, treating source catchment, repair/renovation of existing sources, augmenting discharge from existing sources, recharging sources, creating new sources) which will need convergence among a number of schemes initiated by different departments of the GoI, dealing with water conservation and water management issues. Some of these schemes have a strong sustainability component and may be converged at the district level to optimise their benefits (GoI, 2012).

In a study to assess the sustainability of water sources in the habitations, a sustainability index was calculated comprising of a total of 29 parameters under 5 broad aspects, namely, technological aspects, community and social aspects, financial aspects, water quality aspects and training aspects (GoI, 2010b). The parameters included functionality of source, reliability of source, number of sources with acceptable quality, access to safe water, etc.

Under the National Rural Drinking Water Programme (NRDWP), 20 per cent of the allocation to each state is earmarked for 'sustainability' component, which is provided as 100 per cent grant-in-aid to states, for taking up various works related to sustainability of drinking water sources and schemes (GoI, 2010a). These include constructing structures to use rainwater for storage and recharging of drinking water sources. With traditional water bodies being neglected, the recharge of groundwater sources is an issue. In order to ensure the sustainability of rural water sources, water supply schemes should promote the conservation/recharge of existing sources rather than focusing only on the exploitation of new sources.

Water Demand Management

Water shortage has traditionally been handled from the supply-side management perspective. Gradually, it is realised that a sustainable solution lies in the management of both supply as well as the demand side. India also needs to shift its focus from 'water resources development' to 'water resources management' by restructuring and strengthening existing institutions for better service delivery and resource sustainability and to bypass the 'business as usual' components. In order to make tactical corrections, we need to emphasise on technologies and institutions that are ready to take up the short-term challenges of recycling waste water, storing rainwater, managing storm water, converting saline water for potable use, and enhancing water productivity across diverse uses. Technological innovations are needed to ensure future water supplies (UNESCO, 2009) and to ensure that the country's spending on water management efforts is utilised in a cost-effective manner.

UN-HABITAT in partnership with Water Resource Planning and Conservation (WRP), South Africa, and The Energy and Resources Institute (TERI), India commissioned a Project on Water Demand Management Strategy and Implementation Plan for the cities of Bhopal, Gwalior, Jabalpur and Indore in the state of Madhya Pradesh. This project presents a comprehensive reforms package by developing water demand management strategy and implementation plan for the respective cities involving institutional, financial and technical issues and is aimed at efficiency improvements in the management and utilisation of water. The focus is mainly on the water balancing systems, developing the information-base on geographic information system (GIS) platform, capacity building and approaches for reducing unaccounted-for water for an efficient and effective distribution of available water supply (UN-HABITAT, 2006).

Hydrologic Concepts

The occurrence, storage and distribution of water follow the Hydrologic Cycle and are controlled by the climate, physiography and hydro-geology of the area. Technically, hydrology is defined as the science that treats the waters of the earth: their occurrence, circulation and distribution; their chemical and physical properties; and their reaction to their environment, including their relation to the living things. Hydrology provides a sound basis for study of the interrelationships and interactions between water and its environment in the hydrologic cycle.

Water is the main driving force in a watershed system and almost all other systems, e.g., humans, animals, flora and fauna, soil, and crops are dependent on water for their survival and progress. Watershed is a geo-hydrological unit of area, which drains the entire runoff from the area to a single outlet. The interdependent nature of land and water resources thus necessitates the consideration of watershed as the basic unit in development and planning.

Working on a watershed basis helps in identification of the recharge area or discharge area for surface water as well as groundwater systems. A scientific understanding of the interplay between the two systems, i.e. when an aquifer receives contribution from a stream during high flows or when an aquifer contributes to the stream during low flows, is important in designing sustainable water supply systems. The deterioration of water resources can be contained and the total resources can be properly developed only by adopting a watershed approach.

Hydrological studies provide useful insights into the behaviour of water resources in a project area, improving the capabilities for assessing the quality of the existing water supplies and for identifying alternative and sustainable sources of clean drinking water for the future. Such studies address the problems of freshwater supply and sustainability, the salinity and contamination of groundwater, the evaluation of alternative groundwater resources for potential sustainable exploitation (e.g., Goyal and Gupta, 2003) and assessment of the impact of industrialisation and urbanisation on the quality and quantity of water resources. Hydrology also helps in planning the rejuvenation and management of water bodies (Goyal *et al.*, 2005).

Application of a variety of hydrological methods (hydro-geological, chemical, biological, isotope, etc.) provides invaluable information about the source and movement of water in different environments both above and below the ground, including rivers, lakes and aquifers. Use of modern tools and techniques, such as isotope techniques (RCA, 2007), can also help validate and improve the numerical models based on conventional analytical methods and assist in the overall planning and management of a water resource.

A drinking water supply project usually comprises a source of raw water, a water storage system, a water treatment system and a distribution network of pipelines to reach the demand area (Figure 9.1). The source of water could be groundwater, spring or from a surface water body (e.g., river, lake, pond) or a combination thereof. A sound water resource planning is required to design the whole water supply system.

Figure 9.1

Role of Hydrologic Analyses in a Typical Drinking Water Supply Project

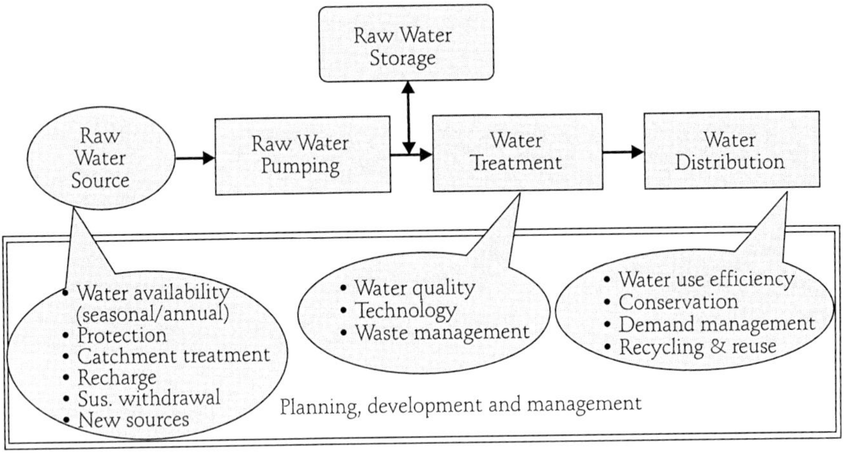

The intake system in such water supply projects could be one of the following:

- directly from a river,
- drawing water from the reservoir behind a dam across a river,
- drawing water from the pool behind a barrage across a river,
- drawing water directly from a lake,
- infiltration wells near a riverbed to draw groundwater, and
- dug wells or tube wells to draw water from groundwater aquifers.

Water Sources

Rain, snow, hail and sleet are precipitated upon the surface of the earth as meteorological water and may be considered as the original source of all the water supplied. Water, as a source of drinking water, occurs as surface water and groundwater. Three aspects should be considered in appraising water resources e.g., the quantity, the quality and the reliability of available water.

Surface Water

Surface water accumulates mainly as a result of direct runoff from precipitation (rain or snow). The amount of available surface water depends largely upon rainfall. When rainfall is limited, the supply of surface water will vary considerably between wet and dry years. Surface water supplies may be further divided into rivers, lakes, ponds and reservoir supplies. Dams are constructed to create artificial storages. Canals or open channels can be constructed to convey surface water to the project sites. The water is also conveyed through pipes by gravity or pumping.

Irrigation water, direct from canals or indirect as seepage, is an important source of drinking water for rural and urban communities in semi-arid countries. Water from seepage sources is of good quality and an important source of clean drinking water. It is important that the use of seepage water for domestic purposes is considered in the planning and management of irrigation systems. In such canal-fed water supply systems, the required quantity of water is stored in the village ponds, which is then passed through Slow Sand Filters and the filtered water is chlorinated and pumped into an overhead tank for distribution through a network of pipes.[1]

Rivers

Many rivers furnish a dependable supply of water. The chief drawback of river water is that it is normally polluted and is quite unfit for drinking without treatment. River water contains dissolved and suspended impurities of all kinds, from surface washings, sewage and municipal wastes, industrial wastes and drainage from agricultural areas.

Groundwater

Groundwater is the cheapest and the most practical mean of providing water to small communities. Groundwater is superior to surface water, because the ground itself provides an effective filtering medium. Groundwater is generally characterised by higher

---

1. For example, see "Sujala Scheme" at *www.byrrajufoundation.org.*

concentrations of dissolved solids, lower levels of colour, higher hardness (as compared to surface water), dissolved gasses and free from microbial contamination. Groundwater pumped from deep wells is largely free from bacteria. The wells require disinfection primarily to prevent the growth of bacteria in the pipes and tanks of water distribution systems.

Springs

Groundwater that flows naturally from the ground is called a spring. Springs may be of two types: shallow springs and deep springs. Shallow springs dry up quickly during summer months, whereas deep springs do not show seasonal fluctuations in the flow of water.

Climate Change and Emergency Scenarios

Climate change is threatening drinking water supplies in two major ways. First, extreme rainfall (higher intensity but in less duration), or lack of it, leads to either high runoff leading to heavier floods (and less infiltration/storage underground) or frequent droughts. Second, rising sea levels will destroy a significant proportion of the freshwater supplies in coastal regions by inundating aquifers with brackish water that is not safe for human consumption. An indirect impact would be due to competing increased demands from other sectors (uses).

In case of extreme events and emergencies such as cyclones, earthquakes, floods, availability of the drinking water supply poses additional challenges, which require advance scientific planning. Also, the growing population and urbanisation scenarios (leading to reduced natural recharge of sources) can be effectively addressed with the help of hydrologic analyses.

Planning for Water Transfer from Adjoining Areas

There are situations when demand for water in an area cannot be met from the existing sources in the catchment area and water transfer is required from adjoining 'water surplus' areas. Tools like GIS provide handy support in analysing such situations (Kumar *et al.,* 2003). GIS is increasingly used in planning drinking water supply schemes all over the world (e.g., Shandas *et al.,* 2003).

GIS platform can also be used to develop decision support systems (DSS) for effective planning and decision making in drinking water supply schemes. One such system is reported by the National Institute of Rural Development, Hyderabad.[2] The software 'Spatial Decision Support System (SDSS) for rural development' has various components related to rural development, including one for Rural Drinking Water Supply and Sanitation Scheme. The software provides a comprehensive report to create sophisticated designs and work estimates at the village level, including preparation of designs, work estimates along with L-Section graphs, cross-sectional diagrams, etc. GIS functionality can also be used to determine minimum cost trajectories of transport pipes.

Water Security and Technology

Competition for water use is drastically increasing and water supply issues can no longer be addressed without considering the use of water for agriculture and industry. Whilst domestic water for drinking is accorded the first priority use in the National Water Policy, it is actually a very small use of water resources. Approximately 80 per cent of the total water resources withdrawn in India are utilised for agriculture and a system is needed that safeguards the relatively small quantities of water required for drinking purposes. To achieve drinking water security, a new approach is required based on a holistic and participatory approach. GPs and village water and sanitation committees need to prepare drinking water security plans which address source sustainability, water quality (both bacteriological and chemical) and O&M issues, as well as provisions for replacement and expansion. The overall approach is to prepare village-wise water budgets (Goyal, 2003), identify how much water is needed for drinking water security and put in place mechanisms to ensure that the required local water resources are protected. In extreme cases, the local resources need to be supplemented with bulk water (GoI, 2011).

---

2. *http://www.nird.org.in/TULIP/SDSS/Rural/Developement/index.html*

Due to the competing demands on scarce water resources from agriculture, industry, domestic use, etc., the management of water is a key issue. To address the problem of water scarcity, there is a need to prepare a holistic blueprint and identify the requirement of water at the village level for these sectors and for drinking water (e.g., how much land in the village needs to be irrigated, how much water is required for industry and the generation of electricity).

Water security can be achieved by the adoption of low-cost technologies (Goyal, 2012). There is a need to build awareness at the grass roots with regard to overdrawing of water, which affects the natural cycles of recharging, as well as issues of water pollution.

The delivery of safe drinking water is most effectively achieved by developing a management system to meet the targets of a water safety plan (WSP) comprising (Howard and Schmoll, 2006):

- An assessment of the water supply system to determine whether the water supply chain (from source through treatment to the point of consumption) as a whole can deliver water that meets the health-based targets.
- Identification and operational monitoring of the control measures in the drinking water supply that is of particular importance in securing drinking water safety.
- Preparation of management plans, documenting the system assessment and monitoring plans and describing actions to be taken in normal operating and incident conditions, including upgrading documentation and communication.

Capacity Building and Awareness Needs

Capacity building for the planning and implementation of water and sanitation (WATSAN) schemes has been identified as a major challenge at various levels, e.g., PRIs, community, Public Health Engineering Department (PHED), etc. There is an excellent network of CCDUs in place, with a specific budget for community activities; these bodies could be used to build the capacity of institutions. Water and sanitation schemes are to be planned and implemented at the community level and involve the complex management of land and water resources, including drainage treatment, land development, liquid and solid waste management, etc. However, at the village level, planning tools and capabilities are not available. There is a need to invest in building local capability for micro level planning and implementation.

There is a need to build awareness among stakeholders so that the issues of sustainable management of water resources are addressed in a scientific manner. Awareness raising activities should also be undertaken in schools, colleges and *anganwadi* centres to target key beneficiaries such as mothers, adolescent girls and children.

Jal Bhagirathi Foundation of Jaipur (India) implemented a five-year project 'Vulnerability reduction through community empowerment and control of water in Marwar region' in 200 villages in Pali, Barmer and Jodhpur districts of the Marwar region in Rajasthan (UNDP India, 2010). Marwar region comprises land of extremes with low rainfall, non-perennial rivers, saline groundwater and extreme temperatures. The project involved community-oriented approach in developing solutions to water problems through construction and conservation of water structures and sources. About 291 water harvesting structures were restored and/or built. A decentralised system of social governance was set up by forming effective and accountable village water user associations called Jal Sabhas (water development group) to plan, implement and manage micro projects. Through contributory Jal Kosh (water development and management fund), communities were able to maintain bank accounts to collect money for carrying out sustenance and maintenance activities.

The project improved the overall availability and access to water for around 300,000 people. The interventions led to the creation of women's Jal Sabhas and Jal Mandals (women's self-help groups) to ensure proactive participation of women in resolving their own water problems. The project facilitated local children to learn about their environment through hands-on experiences in water management systems. Also, the facilitation of small water enterprises in villages as community-owned institutions provided a source of income for the people, especially women, leading to financial security.

Another good example of community-led water management is the Roof Rainwater Harvesting Programme (RRHP) at the Barefoot College of Social Work and Research Centre in Tilonia, Rajasthan. This programme helps communities harvest rainwater in schools, dispensaries and other public places. The approach involves linking together roofs *via* a network of pipes so that any rainwater falling on the surface is channelled into a central cistern, typically built underground. The rainwater is collected, stored and, if needed, a chemical treatment such as chlorination is used to decontaminate the water. The college has seven field centres, each of which works with 20-30 neighbouring villages. The college trains architects and artisans, many of whom may be neo-literate, in construction techniques. The programme has trained more than 1,250 barefoot architects, including women. It has benefited nearly 200,000 people in 18 Indian states, and about 33 million litres of rainwater have been collected every year from the roofs of rural schools and community centres in 492 villages (UNDP India, 2012).

The Ministry of Drinking Water and Sanitation (MDWS, GoI) has identified some institutions of high repute and experience in the water sector to impart training and other capacity building activities for such project implementing agencies (PIA). These key resource centres (KRC) have been assigned the responsibility of capacity building, dissemination of knowledge and information, documentation of best practices, etc. The major objective of the KRC activity of MDWS is to upgrade the knowledge, skills and attitudes of the PHED engineers, PRI representatives, master trainers and other stakeholders operating in different states. It also intends to enhance knowledge and skills about convergence with other related programmes launched at the central and state level. National Institute of Hydrology, Roorkee, has been designated as a national KRC to impart training to selected PHE, PRI and NGO on identified topics of hydrology, such as hydrological cycle principles and watershed concepts, conjunctive use of surface water and groundwater, planning for groundwater recharge, and monitoring and impact assessment of water management projects.

Conclusion

Sustainability of drinking water sources and schemes is a process which facilitates the existing/new drinking water supply projects to provide safe drinking water in adequate quantity—even during distress periods—through conjunctive use of groundwater, surface water and rainwater harvesting. The main aim of providing sustainability in drinking water schemes is that such schemes will not slip back from universal access of safe drinking water to the community throughout the design period of schemes. Adoption of appropriate technology, revival of traditional systems, conjunctive use of surface and groundwater, conservation, rainwater harvesting and recharging of drinking water sources are all considered integral components of achieving drinking water security.

Providing rural population with adequate safe water for drinking, cooking and other domestic basic needs on a sustainable basis requires sharing of resources and skills, including knowledge on the hydrology principles which is vital for long-term sustainability of the water availability (both in quantity and quality). This chapter outlines the role of hydrology in drinking water supply schemes, especially addressing the issue of source sustainability.

Way Forward

With the increasing need for sustainability in the drinking water supply systems, made even more complex due to the impacts of climate change, the role of hydrology needs to be understood and adequately incorporated while designing and implementing such schemes in India. This knowledge has duly enriched in the country over the last three decades. Application of advanced analysis and modelling techniques as well as technology, together with the capacity building of the users and sensitisation of the stakeholders, has vast potential of playing a leading role in ensuring sustainability in the water supply schemes. Hydrologists, on their part, should strive to demystify the complex procedures involving equations, formulations, modelling, etc., and present the hydrologic concepts in user friendly forms for effective use by the community.

References

Bhandari, Betman and Miriam Grant (2007). "User Satisfaction and Sustainability of Drinking Water Schemes in Rural Communities of Nepal". *Sustainability: Science, Practice & Policy* 3(1) (*http://ejournal.nbii.org.*)

GoI (2010a). *National Rural Drinking Water Programme.* Rajiv Gandhi National Drinking Water Mission, Ministry of Rural Development, Department of Drinking Water Supply. pp.90.

———. (2010b). *Evaluation Study on Rajiv Gandhi National Drinking Water Mission (RGNDWM).* New Delhi: Programme Evaluation Organisation, Planning Commission. pp.97.

———. (2011). *Towards Drinking Water Security in India: Lessons from the Field.* Ministry of Rural Development, Department of Drinking Water and Sanitation, and Water and Sanitation Program. New Delhi: The World Bank.

———. (2012). *Convergence for Sustainability. Sustainable Drinking Water Supply—Sanitation for All.* Ministry of Rural Development, Department of Drinking Water Supply, Rajiv Gandhi National Drinking Water Mission. pp.28.

Goyal, V.C. (2003). "Assessment of Rainwater in the Kandi Belt of Jammu Siwaliks". *Hydrology Journal* (IAH) 26(1-2).

———. (2012). "Science and Technology for Water Security in Indian context", *Plenary Lecture* at the 99th Indian Science Congress, Bhubaneswar. January 3-7.

Goyal, V.C. and P.K. Gupta (2003). "Resistivity Soundings for Hydrogeologic Evaluation in Piedmont Region of Jammu Siwaliks (India)", *International Symposium on the Application of Geophysics to Engineering and Environmental Problems* (SAGEEP 2003), San Antonio, Texas (USA). April 6-10.

Goyal, V.C., Vijay Kumar, S.P. Rai and Omkar Singh (2005). "Hydrological Aspects of Rainwater Harvesting for a Village Pond in the Kandi Belt of Jammu Region (India)", *Int. Conference on Hydrological Perspectives for Sustainable Development* (HYPESD-2005). Roorkee: Department of Hydrology, IIT Roorkee. February 23-25.

Howard, G. and O. Schmoll (2006). "Water Safety Plans: Risk Management Approaches for the Delivery of Safe Drinking-water from Groundwater Sources", in O. Schmoll, G. Howard, J. Chilton and I. Chorus (eds.), *Protecting Groundwater for Health: Managing the Quality of Drinking-water Sources.* World Health Organisation. London, UK: IWA Publishing.

James, A.J. (2003). "Linking Water Supply and Rural Enterprise: Issues and Illustrations from India", *International Symposium on Water, Poverty and Productive Uses of Water at the Household Level.* January 21-23. Muldersdrift, South Africa.

Kenel, P.P., P.E. Black and Veatch (2005). *Practices to Promote Sustainable Water Sources.* WEFTEC. pp.5650-56.

Kumar, Sudhir, V.C. Goyal and Lekh Raj (2003). "GIS-based Study for Spring Water Management in a Hilly Watershed", *National Seminar on Hydrology of Small Watersheds*, Trivandrum. May 6-7.

Reed, B.J. (2001). *Working Together—The Sharing of Water and Sanitation Support Services for Small Towns and Villages.* Leicestershire, UK: Water, Engineering and Development Centre, Loughborough University,

RCA (2007). "Isotope Hydrology helps find Water Fit to Drink: RCA Success Story 2007". Regional Cooperative Agreement for Research Development & Training Related to Nuclear Science & Technology for Asia and the Pacific, Korea (*www.rcaro.org*).

SCBD (2010). *Drinking Water, Biodiversity and Development: A Good Practice Guide.* Montreal: Secretariat of the Convention on Biological Diversity. pp.41.

Shandas, Vivek, M. Alberti, J. Gibson, S. Moddemeyer and J.M. Irons (2003). "A GIS Based Water Demand Analysis for Municipal Application". *Map India Conference 2003*, New Delhi.

UNDP India (2010). *Good Practices in Water Security—Ideas for Praxis*. New Delhi: United Nations Development Programme.

———. (2012). *Greening Rural Development in India* (Volume 1). New Delhi: United Nations Development Programme.

UNESCO (2009). "Water in a Changing World", *The United Nations World Water Development Report* 3 (WWDR 020409.indd).

UN-HABITAT (2006). "Water Demand Management Strategy and Implementation Plan for Gwalior", Water for Asian Cities Programme (Project UDAY). United Nations Human Settlements Programme (UN-HABITAT), Nairobi, Kenya.

World Bank (2010). *Deep Wells and Prudence: Towards Pragmatic Action for Addressing Groundwater Overexploitation in India*. Washington, D.C.: The International Bank for Reconstruction and Development/The World Bank.

10

BRECHT MOMMEN and PRAVIN MORE

Water Service Sustainability

A Study of Odisha

Introduction

India has made remarkable progress in providing access to safe water sources, but is also facing challenges related to service sustainability. The paper analyses the sector, from a service perspective, and uses Odisha as a case study. Based on the analysis, root causes undermining service delivery are identified and recommendations given.

Box 10.1

Root Cause of Limited O&M

Drinking water supply schemes have suffered from poor upkeep. Responsibility for operation and maintenance (O&M) of water supply schemes lies with the panchayati raj institutions (PRIs), but in many states, this responsibility is poorly defined and not supported by transfer of adequate funds and trained manpower to the PRIs. PRIs and village water and sanitation committees (VWSCs) are not willing to take over completed schemes in which they were not involved at the planning and implementation stages. Inadequate water resource investigation, improper design, poor construction, substandard materials and workmanship, and lack of preventive maintenance also lead to rapid deterioration of water supply schemes.

Source: Planning Commission, 2011.

Background

Impressive achievements have been made under the National Rural Drinking Water Programme (NRDWP) and its predecessors in increasing access of rural households to improved water sources. The recent census data analysis indicates that the Ministry of Drinking Water and Sanitation has ensured that 96 per cent of the households in India are covered by a safe source: piped water scheme (30.8%), hand pump (51.9%) and well (13.3%).[1]

In line with the 73rd Constitutional Amendment, the NRDWP framework supports the transfer of management and financial responsibilities of the rural water supply (RWS) service delivery to the lowest level of PRI and in particular the VWSC. The transfer includes the management and financial autonomy to the PRIs and VWSC [Government of India (GoI), 2011].

Despite the achievements in hardware creation, challenges have been observed related to water service delivery. It is estimated that currently 40 per cent of the groundwater sources in India are microbiologically contaminated. Much of this contamination is preventable by proper O&M of water sources coupled with safe sanitation practices.[2] Root causes of limited O&M have been identified by the Planning Commission of the GoI, related to the limited capacities to maintain and the limited involvement in the planning stages of water source development of PRIs (Planning Commission, 2011).

This paper investigates this matter deeply and uses rural Odisha as a case study and explores potential solutions.

Odisha Case Study

Odisha is located on the east coast of India; its neighbouring states are West Bengal, Jharkhand, Chhattisgarh and Andhra Pradesh. The state has a population of about 42 million people. For administrative purposes, the state has been divided into 30 districts and 314 development blocks.

1. Census 2011, Government of India.
2. Government of India, Ministry of Rural Development, Department of Drinking Water and Sanitation; No w-11042/01/2011/water.

Figure 10.1

Odisha, Administrative Divisions

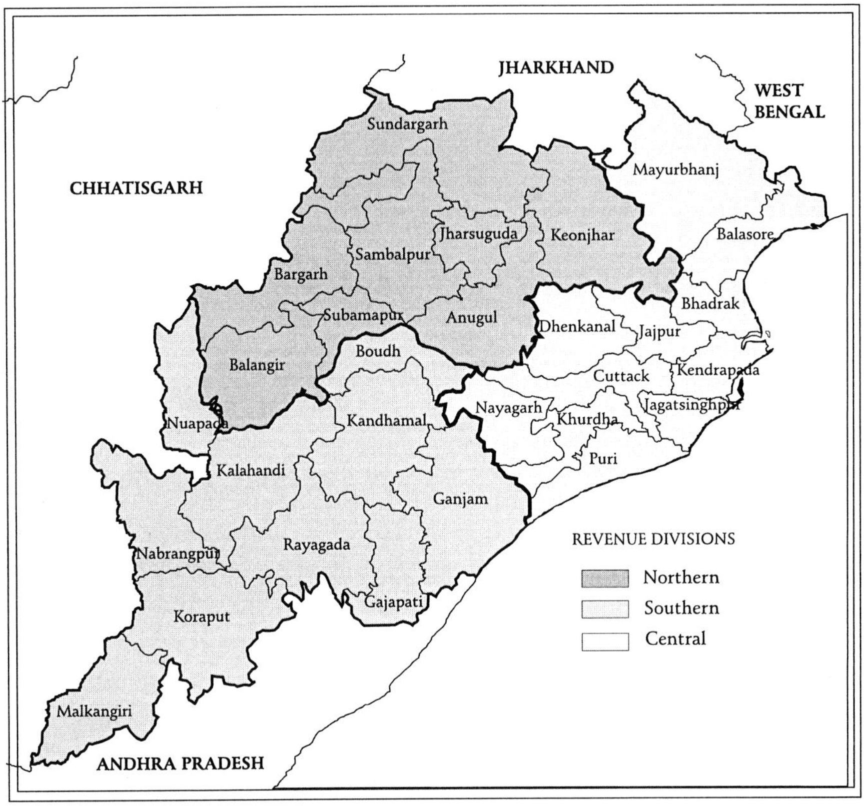

Odisha has made substantial progress in providing access to safe water. Over a timeframe of 30 years, the access to 'other sources' (springs, streams, ponds) and wells has reduced dramatically and replaced by hand pumps. The progress in piped water supply has been minimal.[3] Gradually, the focus of Rural Water Supply and Sanitation (RWSS) Organisation has shifted towards climbing the water service ladder and providing access to piped water supply.

3. Census 1981-2011.

Figure 10.2

Progress of Water Supply in Odisha, 1981-2011

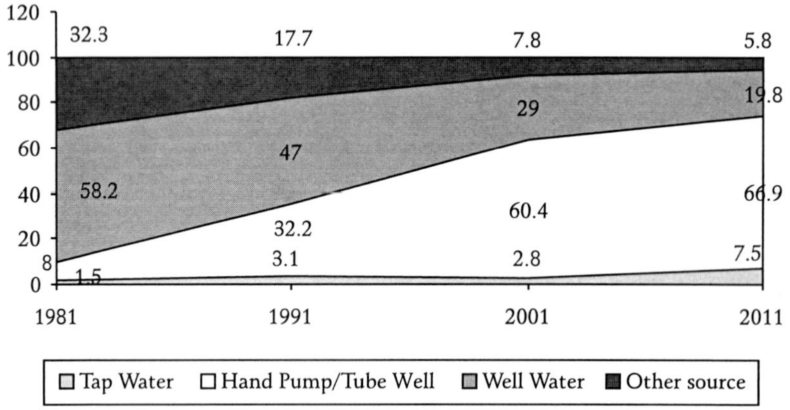

Source: Census.

Despite these significant achievements of the state, access of households to improved water sources is accompanied with several challenges that compromise the reliability of service. This is evident from the fact that more than a third of the rural households are forced to access a second source to meet their drinking water needs.[4] In villages covered with piped water supply, these pumps and wells have remained crucial, as most piped water schemes in Odisha lack an overhead tank and thus water is only available during pumping hours, which is generally in the morning and evening hours.

The concerns of lack of O&M and its effect on water quality are strongly evident in Odisha. Water quality testing results by the district level laboratory in the district of Koraput indicate that 80 per cent samples out of the 1,593 samples taken in 2009 and 2010 has faecal streptococci contamination (Cronin *et al.*, 2009). The water quality sampling in 2010 also included a sanitary surveillance. It was found that more than 70 per cent of the sources had five or more risks present. The analysis of accompanying sanitary surveillance data with the water quality testing emphasised a strong corelation between biological contamination and the insanitary conditions around the water sources (Ibid).

4. National Sample Survey (NSS), Round 65, 2009.

Figure 10.3

Water Quality Test Results, Koraput District, 2010

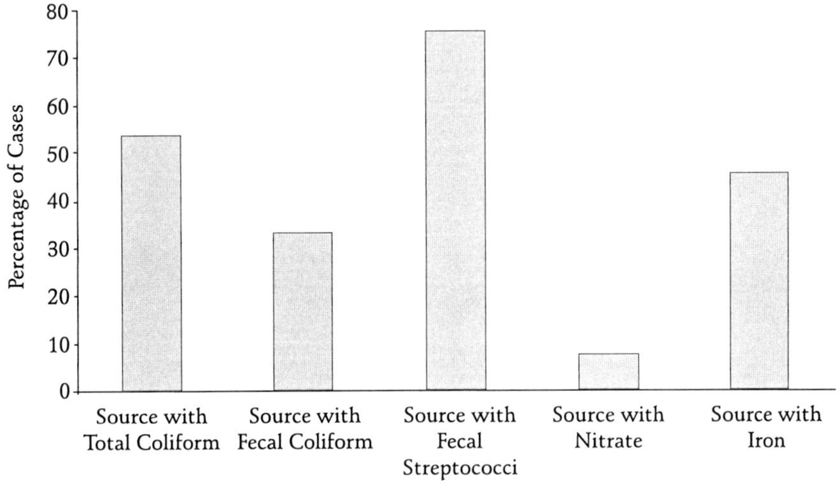

The issues described in the approach paper give a clear indication that the lack of O&M has its roots in the sectoral approaches. This paper makes an attempt to understand the causes which allow this situation to prevail *via* systematic analysis of the elements as given in Figure 10.4.

Figure 10.4

Water as a Service

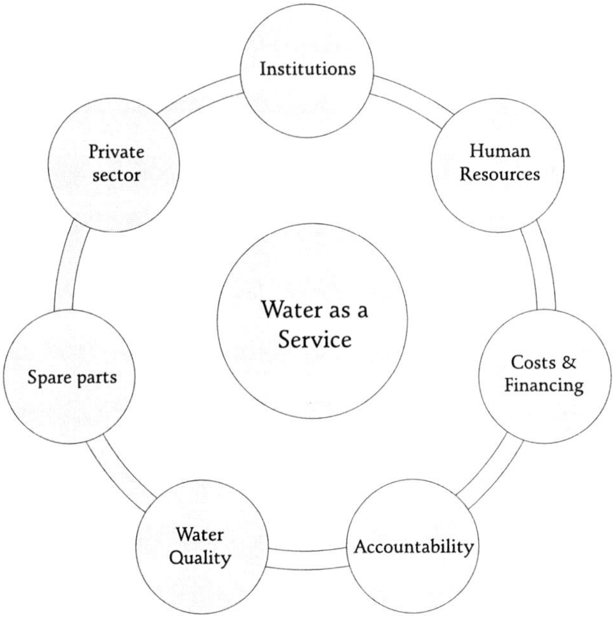

Figure 10.5

Sanitary Conditions

Source: Image Courtesy: Pravin More, 2010.

Institutions

The national level framework gives guidance on the sector institutions and their roles. However, 'water supply' being the state subject, the responsibility for developing state-specific framework with clear roles and responsibilities remains with the states. In the context of Odisha, the sectoral responsibilities are distributed among various institutions as visualised in Figure 10.6. The consumer at the heart of the RWS sector is supported by multiple institutions.

Figure 10.6

Institutional Setup

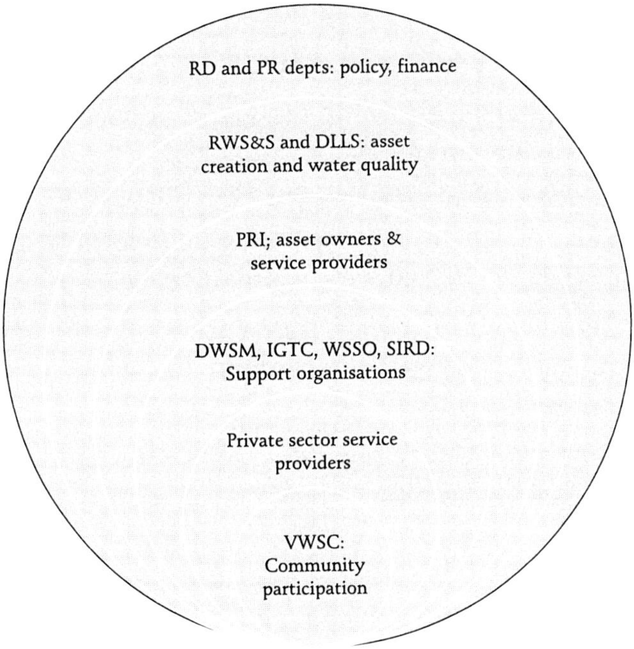

- The Departments of Rural Development and Panchayati Raj Development: Policy guidance and financing sectoral investments.
- The RWS&S as the asset creator and, under its umbrella, the District Level Laboratories (DLL) which undertakes water quality testing.
- The PRIs as the asset owners, responsible for O&M and supplies of spares through block development office.
- The support organisations as: District Water Sanitation Missions (DWSM), Water Sanitation Support Organisation (WSSO), Indira Gandhi Training Centre (IGTC) and State Institute for Rural Development (SIRD) as support and training organisations.

- The private sector and self-employed mechanics (SEMs) as the service providers for construction of hardware, supplies of spares and basic maintenance.
- The VWSCs, to ensure community participation.

In order to ensure sustainable water service delivery to its consumer, it is important that there is high level of coordination and synchronised functioning among these institutions.

Generally, the O&M of sources is broadly possible singularly or *via* combination of the following approaches:

- *Community managed:* Wherein the assets remain common property resource of the community which in turn takes responsibility for O&M.
- *Community with Private Sector:* Wherein community retains the asset ownership and outsources asset maintenance to a private agency.
- *Public Sector/Government Agency based:* Wherein government agency/department owns and takes responsibility for repair and maintenance.
- *Public-Private Partnership (PPP)*: wherein the asset ownership remains with the government agency and maintenance is outsourced to a private agency.

In Odisha, the policy framework outlines a combination of community (with private sector) and public sector based approach has been adopted for O&M management of RWS services. The responsibility for asset formation rests with the government. After commissioning, asset ownership is transferred to the PRIs—which is a community-based constitutional body. The PRIs then carry out O&M of infrastructure through a SEM and spares are purchased by the government.

As per the decentralisation process, the executive engineers (at division level) and assistant engineers (at sub-division level) under RWS&S are accountable to the *zila parishad* (ZP), the highest tier of Panchayati Raj system. This has been communicated in the Government of Odisha circular in 2003.[5] However, the implementation in true spirit is not strongly evident. RWS&S has remained an independent asset creation authority, with limited involvement of the PRIs. The PRIs, which are supposed to own, operate and maintain the assets and ensure service delivery, severely lack the capacity to monitor/verify quality of installed hardware. Anecdotal evidence suggests that RWS&S has remained the creator and the caretaker of rural water supply, for example, chlorination of all sources before monsoon is led by RWS&S and not by PRI.

In 2002, State and District Water and Sanitation Missions (SWSM and DWSM) were formed to ensure community-oriented planning and implementation of RWS schemes. The mission, however, retained the conventional functioning style and character—unwilling to be the facilitator. This paradox is highlighted by the fact that at the district level, the two positions—Executive Engineer of RWS&S and the Member Secretary of DWSM—are held by one individual.

The collaboration mechanisms with SEMs are also complex. Over time, the number of water sources has increased while the number of SEMs has not increased. Over the timeframe 2002-2010, the ratio of number of sources covered per SEM has increased from 29 to 40. This significantly increases the workload per SEM. One of the possible causes is reportedly related to the fee given to SEMs for regular maintenance. RWS&S field functionaries and the PRI members argue that the fee is not sufficient to sustain sufficient number of SEMs. This is especially applicable in remote thinly populated areas with low water source densities, creating larger distances for SEMs to cover water sources dispersed over larger number of habitations. Increasing the fee for SEMs has been advocated for but awaits approval from the State level. If decentralisation had been implemented in the true spirit, could this decision not been made at the block or GP level?

There are also challenges around the role of communities in the VWSC. According to the NRDWP framework, it is expected that households take 'ownership' of the sources. However, in Odisha this

5. Government of Odisha letter; Npl-PS-2/2003-6886/ PS dated 04/07/2003.

'community ownership' is not being felt. Most communities feel that these water sources are government property and therefore, it is not the communities' responsibility to look after the sources. There is validity in this perception as per the adopted model in Odisha; hardware is created by RWS&S and assets handed over to PRI with a budget line for O&M. Given this adopted model, can we then expect 'community ownership'?

These are some key institutional gaps. Further investigation is needed to understand the full spectrum of the complexities of the institutional structure and the alignment of individual agencies with their desired roles and responsibilities. It is recommended to position the consumers—the households—at the heart of the structure and to design an institutional set-up to ensure sustainable water service delivery to its consumers.

Human Resources

Human resource (HR) is one of the key contributing factors in ensuring functionality of any system; this is the same for delivering water as a service. Limited capacities of sectoral institutions is a challenge across India, and the Planning Commission identifies the limited trained HRs of PRIs as one of the root causes of the poor upkeep of drinking water sources.

In Odisha, this gap has been reiterated during stakeholder consultations; the gap in HR capacities exists both within PRIs and RWSS organisation. HR capacity limitations are not only in terms of availability of HRs, but also in skill sets. A recent assessment of training needs for executive engineers revealed that there is no specific job description for executive engineers, only a table of authorities. This severely challenges the overall HR management such as performance reviews, recruitment and promotions. It is believed that more posts in the system have no specific job description. Being an engineering organisation, traditionally, the RWSS lacked community mobilisations skills, which are equally important.

Beside the gaps in knowledge and skills of those people in place, there is a persistent gap in filling all the approved posts within RWS&S. As on May 2012, a total of 16 per cent of the posts were vacant. This proportion is observed to be the same in the year 2009-10 [*Annual Report* of Rural Development Department (RDD), 2009-10], when the position of JE-II was not created. It is clear that vacancies undermine the performance of RWS&S.

Figure 10.7

RWSS Filled and Vacant Positions (May 2012)

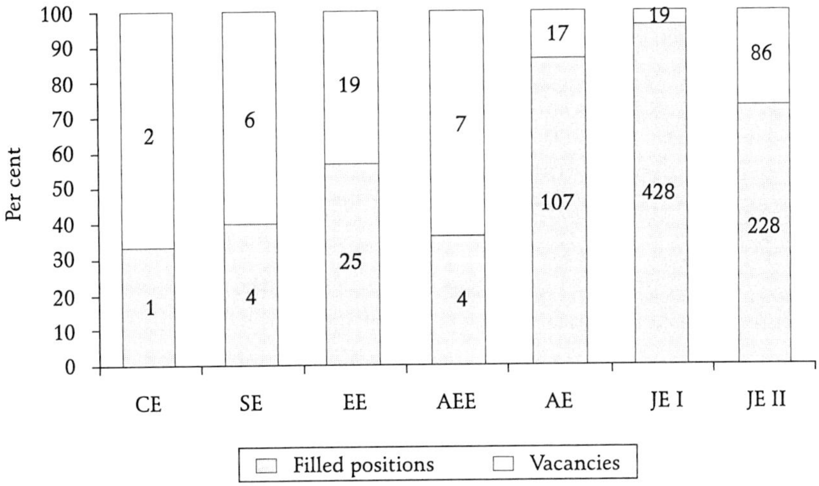

Note: CE–Chief Engineer; SE-Superintending Engineer; EE-Executive Engineer; AEE-Assistant Executive Engineer; AE-Assistant Engineer; JE-I-Junior Engineer, RWSS; JE-II-Junior Engineer, O&M of Rural Water Supply.

Source: *www.rdodisha.in.*

Capacity building in Odisha is being spearheaded by two training institutes—SIRD aimed at building the capacities of PRIs and IGTC aimed at building the capacities of the water and sanitation sector professionals. Besides these two training institutes, external agencies, such as the key resource centres (KRC) engaged by the GoI, provides trainings for RWSS staff. The SIRD has a well-established curriculum; however, training the large number of PRI members presents a serious challenge due to limited capacities. The review of training manual—sections relating to RWSS sector—indicates that the mode of training is rather instructive with a detailed description of schemes and programmes. However, it does not seem to address the issue of how to translate the benefits of these schemes and programmes for the rural populations. There is a weak follow-up mechanism after the trainings, and it is unclear to what extent the trainings enable PRI members in carrying out their duties and what the gaps are.

The other trainings organised by the IGTC and external agencies often miss out the identification of needs of the target group and the end measurement. The other challenge being that the interpretation of the term 'capacity building' is often limited to 'training.' The capacity building measures thus miss the supportive actions necessary to create an enabling environment. It is therefore unclear whether these trainings deliver envisioned objectives.

It is recommended that the RWSS sector adopts a systematic approach to capacity building. Some specific measures could include definition of job descriptions for various positions, tailored training and capacity-building inputs based on identified needs, adoption of results-based performance monitoring framework, identification of systemic barriers and creating an enabling environment.

Costs and Financing

A basic principle for sustainability is that the O&M costs should be 100 per cent financed. The financing can be obtained from user charges by consumers and government financing systems of RWS&S and PRI.

In Odisha, the cost of operating and maintaining RWS systems are financed primarily by the government—through RWS&S and PRIs. User fees are only collected from household connections with a piped water supply (PWS), which is a very small proportion of the 7.5 per cent households accessing the PWS as most of the PWS systems have only public stand post and lack household connections. The RWS&S is responsible for infrastructure creation, augmentation, renovations and water quality monitoring, while PRIs are responsible for O&M. This role division between government departments is strange, as the capital and O&M costs are interconnected and strongly influence each other. Choices made in infrastructure creation affect the expenditures in O&M, while lack of regular O&M inflates the costs of renovations.

As the sector is moving away from the project model, towards a sustainable service delivery model, it is advisable to calculate the total life cycle costs for all newly created hardware. Dividing these costs into two departments' budgets discourages the life cycle approach. It would therefore be advisable to have the overall budget with one department, as this would support the overall life cycle costs approach as it would be carried by one department.

Concerning the existing infrastructure, the O&M costs are financed through PRIs based on Finance Commission recommendations and sectoral outlays. The current budget practice does not encourage bottom up estimation based on the scheme/village/*panchayat*-specific O&M requirements. Costing estimates are based on indicative spare part replacements estimates developed in the 1980s. This is a flat rate per year, which doesn't include critical elements such as the lifespan of the source, the individual load intensities and depth of the groundwater. Additional costing factors such as the costs of transportation and human resource, overheads and monitoring are also not included. This lack of detailed breakdown affects the financing side. Financing of O&M costs are thus ill informed. Anecdotal evidence suggests that sufficient funds are not allocated for O&M.

It is therefore recommended to develop a comprehensive overview of the expected maintenance costs of RWS service delivery over a longer timeframe—e.g., the next 10 years—which forms the basis for financing the expenditures. For hand pumps, a prediction of the cost should include: manpower requirements, spares, transport, repairs, monitoring and overheads. This prediction could be based on basic parameters such as year of construction, working load and groundwater level. For piped schemes, the prediction should include: energy costs related to pumping, the replacements of the electric-mechanical parts and overhead costs. Based on these breakdowns, a comprehensive picture can be displayed in a matrix form, per sources, per GP, block, district and state over a longer timeframe.

Table 10.1

Indicative Format for Recording Estimated Annual O&M Costs

Location	*Source No.*	*Type*	*Depth*	*Population*	*Construction Year*	*O&M Costs per Year*				
						2012	*2013*	*2014*	*2015*	*2016*
GP x	0001	HP	20	200	1999					
	0002	HP	25	400	2003					
	0003	WL	10	50	1997					
	0004	HP	30	150	2010					
Total GP x	800									

Note: HP, hand pump; WL, well.

This overview can be compiled for the blocks and districts. This approach would support the planning, budgeting and monitoring and makes the whole process of costs, financing and expenditures more transparent.

Figure 10.8

Accountability

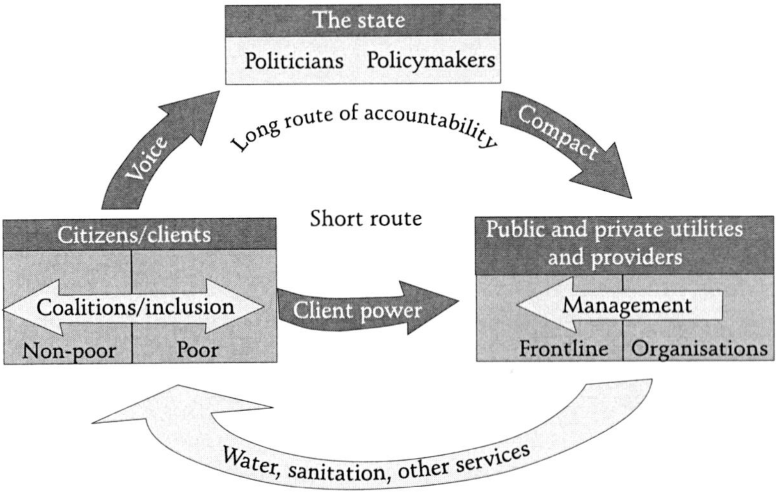

Source: WDR, 2004.

Accountability

Accountability is one of the fundamentals of good governance. As the primary role of a government is to deliver services to its citizens, holding a government/service provider accountable can improve the quality and sustainability of the delivered service, and improve the responsiveness of a government towards its citizens.

The *World Development Report* (WDR) (2004) argues that there are two routes of accountability in the service delivery of water and sanitation services: the long route and short route. The long route is indirect, layered, tedious and inherently slow and complex. In this, the collective demand of the population has to be heard by the politicians/policy makers, who in turn need to make appropriate policies/adequate provisions for adoption by the service providers. Some of the key challenges of the long route of accountability are: (i) elections take place once in five years; (ii) water supply may or may not be the priority for politicians and policy makers; (iii) RWS being a monopoly market, establishing accountability is often an impossible task.

In the short route, accountability is established directly between the service provider and clients. The National Rural Water Sector framework has a section on community monitoring and social audits. The section specifically mentions monitoring of consumer satisfaction (GoI, 2011). The short route is therefore recommended to be strengthened, as it facilitates establishment of a direct linkage between the service providers and consumers.

Odisha has made partial progress in establishing short accountability route by the decentralisation of the PRIs and the VWSCs and the initiation of a Sanjog helpline. O&M of RWS systems has been decentralised and thus brought closer to the clients. However, the prevailing lack of clarity on the roles between RWS&S and PRIs also affects the lack of accountability towards the consumers. The complexity is further compounded by the fact that the funding for both asset creation and the maintenance (including supply of spares) is made available by the government. Thus, the accountability of these institutions remains towards the financer and not towards its consumers. Additionally, the state has put in place a complaint registration system—Sanjog helpline—wherein users can register complaints. However, this system does not include a feedback loop. The importance of the feedback loop has also been highlighted by the Secretary of the Ministry of Drinking Water and Sanitation, GoI: "the current redressal system is been degraded to a "complaint" mechanism."[6] Moreover, anecdotal evidence suggests that the overall majority of the complaints are not being filed through this helpline, but by an informal route of communication, with limited accountability.

Key recommendations towards improving accountability of the water service providers towards its consumers include: (i) supporting the devolution of water service delivery to PRIs with adequate capacity building to promote asset ownership and clear delineation of responsibility, (ii) strengthen redressal mechanisms (such as Sanjog helpline) with central focus on client satisfaction and feedback loop, (iii) institute mechanisms for client satisfaction based ranking of *gram panchayats* (GPs)/blocks and districts; this can help in creating a sense of competition and 'wanting do to better' can be promoted.

Water Quality Monitoring

Water quality monitoring is an integral part of the NRDWP framework, as the commitment to ensure safe water availability at the household level is the core of the service delivery. In Odisha, water quality monitoring is being taken care by the RWS&S organisation. District level laboratories (DLLs) have been set up across the state. Conceptually, the test results should be shared with PRI at block/GP level, who can develop water safety plans with the community. Regular testing should support the development of water safety plans as a tool to validate to what extent these plans have been effective. The results should therefore be shared with PRIs and users to ensure a constant information loop.

6. Secretary of GoI, Ministry of Drinking Water and Sanitation, during the state NRDWP TSC review meeting on January 19, 2012.

Figure 10.9
Water Quality Monitoring Loop

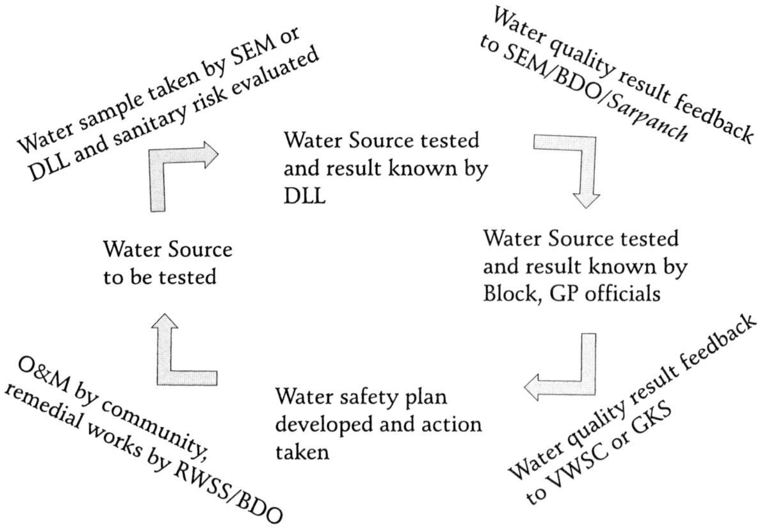

Source: Cronin *et al*. (2009).

The test results from the district laboratory are being uploaded on the website of the department. A quick scan of results in Odisha however shows primarily chemical contamination, without further analysis and no results related to biological contamination were found.[7] In case chemical contamination is identified, this is being communicated towards RWS&S Organisation to initiate corrective actions, which may also include creation of new hardware. In case microbial contamination is tested, it is doubtful whether the feedback loop to the PRI actually takes place. As the results are not communicated, there is also a lack of corrective action by the PRIs or VWSCs.

The water quality monitoring needs substantial strengthening by instituting systematic sanitary surveillance, promoting the role of PRIs in water safety through development, and follow-up on water security and safety mechanisms. Further strengthening of water quality monitoring can be achieved by supporting

7. *http://indiawater.gov.in/imisreports/nrdwpmain.aspx.*

independent assessments, timely inputs for and corrective actions by the PRIs.

Spare Part Supply Lines and Innovative Service Models

Spares and the supply lines are a crucial component for O&M. Lack of access to spares or weak supply lines affects the O&M of assets and thus access to water. As a principle for effective O&M, access to spares should not cause delays in the repairs and spares should be available at the lowest available costs. The NRDWP framework doesn't provide guidance here, as it is considered to be a state subject.

In Odisha, spare parts are being purchased at the block level by the purchase committee chaired by the Block Development Officer (BDO), using the approved rates by the Chief Engineer from four different empanelled suppliers. SEMs can receive minor spare parts such as washers and bolts in advance during monthly meetings, while other parts are being kept at the block level.

Studies indicate that the supply of spare parts is one of the major problem and several SEMs reported not having received the spares at these monthly meetings. This situation is due to delays in fund disbursement and also due to delays on the part of suppliers (Srivastava and Cronin, n.d.). Additionally, it is not very clear which types of spares can be kept by the SEM and which ought to be stored at the block, while this can have large implications.

Given the crucial role of spare supply lines in service delivery, it is recommended to review this practice and recommend supply routes that promote more autonomy to the GPs, increase cost effectiveness and reduce system downtime. The required maintenance prediction over a longer timeframe could prove to be instrumental here. An example of a strategy aiming at reducing the downtime could be preventive replacements per predefined routine maintenance schedule over expected service life of assets, with parts with a high turnover, low costs and high predictability of the spares requirement, for example, the washers. Another way could be to explore modality of PPP.

The Planning Commission emphasises the need to invest in the (water) infrastructure, as this is critical for sustaining and accelerating growth. The approach paper encourages models of PPP as a tool to increase the sector investments. Public investment in infrastructure will have to bear a large part of the brunt of meeting infrastructure needs in backward and remote areas to expand the much needed public services. Since resource constraints will continue to limit public investment in infrastructure, in other areas, PPP-based development needs to be encouraged wherever feasible with appropriate regulation and concern for equity (Planning Commission, 2011). This creates opportunities to explore new models for service delivery in rural areas.

Globally there are several experiences on using a PPP 'service contract' model. These types of service contracts have incorporated a part of the lifecycle costs in a contract. Some examples of the service contracts are: hand pump leasing schemes, total warranty scheme, water assurance schemes (World Bank, 2006). These contracts are focused on ensuring service provision of water. The key deliverable of these contracts is therefore sustainable safe water availability. These types of contracts are mostly used when constructing a new water source. Instead of awarding a contract to construct a water source, the contract is awarded to ensure safe water provision for an agreed timeframe. Thus, the construction costs and the total O&M costs are included into one contract and the deliverable is access to safe water. The main advantage in these contracts is the reflection of the lifecycle costs and not only the investment costs. Secondary benefit is that it encourages innovative cost-effective approaches, as the service providers engage in a competitive bid for the lifecycle costs without prescribed technical solutions, creating space for innovative cost-effective approaches. Clustered contracts incorporating surveys, construction and O&M as a combined package of service into one contract have shown promising results in Africa and have been recommended as a model to achieve twin benefits of cost reduction and reliable service.[8]

8. African Hand pump Market and Supply Division Mapping Study, Delta Partnership, July 2009 on behalf of UNICEF WASH Section, New York.

A global comparative study on water services and the private sector in developing countries reveals that PPP contracts for water show mixed results. The study concludes that increased investments are not the right justification for a PPP in the water sector as companies are reluctant to invest where the profits are low and the perceived risks are high. The gains found were most in the improved efficiency. The study concludes that there is no optimal model and that for every setting, a localised model is the way forward. As the topic is highly politicised, political dialogue is the first conditionality. Stakeholder alliance building and creating a mutual understanding between government, public and the private sector is another important conditionality. In rural water supply, the support towards local entrepreneurs has been identified as another important conditionality for an effective PPP model.[9]

Some innovative service models have been piloted in India. In Uttar Pradesh, two approaches have been piloted. In the district of Lalitpur, a hand pump repair centre has been established at block level with support of UNICEF.[10] In Mohoba district, technical centres have been established with support of Water Aid. The centres provide service to repair and supply spares to *panchayats*. In Gujarat, multi-village piped water schemes have been handed over to the Gujarat Water Supply and Sewage Board (GWSSB), which have outsourced O&M to a third party.[11]

In Odisha, these models have not been tested. Based on global experiences, it is believed that such models have the potential to reduce costs and improve service delivery. It is recommended to pilot modalities within India to gain understanding on how these will function in different contexts. The main challenge towards such service contracts is that it limits the influence of the government service providers compared to the conventional approach. Politicians, RWS&S organisation and PRI would have to change their role from implementers to facilitators, supervisors and

9. Water services and the private sector in developing countries; Comparative perceptions and discussion Dynamics, Aymeric Blanc and Sarah Botton, Agence Française de Dévelopement, Paris, March 2012

10. Nair (2010).

11. Jacob and Lala (2011).

regulators in service delivery. The process of changing roles can be difficult, as observed from the sector reforms, to transform RWS&S from independent hardware creators to facilitators. To imitate this process, political dialogue and stakeholder consultations with the private sector should be initiated to build trust and develop some workable models which could be tested.

Conclusions and Recommendations

Impressive achievements have been made in increasing access to safe water sources, as the sector has accelerated in the infrastructure creation to provide access to safe water to 95 per cent of the households. However, there are major challenges around water quality and O&M. The problems in maintaining water quality and O&M are expected to increase as more (complex) infrastructure is created.

The institutional set-up needs a thorough investigation. There is a gap between the desired and the practice of the institutional arrangements. The lack of role clarity between the water service asset creators and service providers affects their ability to deliver sustainable services to the rural households. Limited human resource capacity and the lack of a systematic approach toward human resource management weakens the performance of the institutions.

The transition from the project model to a service model is not being supported by the budget division between hardware creation and regular O&M. To fully embrace the service and lifecycle approach, one budget carried by one department would be recommended. Current O&M plans and budgets are developed based on incomprehensive and outdated estimates, leading to uninformed decision making. Developing comprehensive cost predictions over a longer timeframe would support planning, budgeting and monitoring of the expenditures and would support transparent bottom up planning. The short route of accountability, between service providers and clients, requires further strengthening to improve service delivery. Measuring client satisfaction could be one of the tools to measure and improve service delivery.

The link between water quality monitoring and improving O&M has been found to be weak. Repositioning the tool of water quality monitoring closer to PRIs could improve O&M and thus the water quality.

The supply chains of spares have space for improvements, as they are crucial for O&M. Exploring spare part supply routes that promote more autonomy to the GPs, increase cost effectiveness and reduce system downtime is recommended. Global experience suggests that innovative service delivery models using a PPP approach have the potential to enhance the efficiency of service provision. Stakeholder dialogue between service providers, political leadership and private sector could identify potential models, while sharing these experiences at a national level could further enhance sector learning.

References

Cronin, A.A., M. Ohikata and R.P. Mishra (2009). *Orissa Water Quality Monitoring: Situational Analysis and Potential Ways Forward*. UNICEF India unpublished report.

Government of India (2011). *National Rural Drinking Water Program: Movement towards Ensuring People's Drinking Water Security in Rural India*. Framework for Implementation. Department of Drinking Water Supply, Ministry of Rural Development, Government of India. April.

Jacob, Nitya and Sunetra Lala (2011). *Consolidated Reply: Ways to Improve O&M of Rural Water Supply Schemes—Experiences, Referrals*. New Delhi: United Nations Solution Exchange, Water Community of Practice.

Nair, Padmaja (2010). *Public- Private Participation Model for Hand Pump Maintenance in Lalitpur: A Perspective on Impact and Scale Up*. Draft Report. New Delhi: UNICEF.

Planning Commission (2011). *Faster, Sustainable and more Inclusive Growth: An Approach to the Twelfth Five Year Plan*. Planning Commission. Government of India. October. pp.9 & 62.

Srivastava, P. and A. Cronin (n.d.). *Strengthening Rural Ware Supply via the Self Employed Mechanics System in Orissa*. UNICEF (Internal Publication). Bhubneswar.

World Bank (2004). "Making the Services Work for Poor People", *World Development Report*. Washington, DC: World Bank.

——. (2006). *Spare Part Supplies for Hand Pumps in Africa. Success Factors for Sustainability*. Water and Sanitation Program—Africa, October.

Disclaimer: Both the authors are employed by UNICEF India Country Office. However, the views expressed herein are those of the authors and do not necessarily reflect the views of UNICEF or those of the United Nations.

11

A.K. MISHRA, B.S. BISHT and
SANJEEV SHARMA

Water Quality Surveillance in Nainital

Introduction

While household sanitation improvement is considered one of humanity's greatest advances of the last century, nearly two in every five individuals still live without improved basic sanitation today. The age-old undignified and unhealthy practice of open defecation continues to contaminate the environment resulting in diarrhea, malnutrition and death. In fact, waterborne illness and poor sanitation lead to death of 1.4 million children yearly worldwide (Water Aid, 2010a). Over the globe, Southern Asia has the worst sanitation record with 58 per cent open defecation rate, and out of this, India is by far the largest contributor. One of two open defecators in the world is Indian (WHO and UNICEF, 2010). Still, in 2008, approximately 69 per cent of Indians, mostly in rural areas, lived without improved basic sanitation resulting in serious consequences. Half a million Indians die from water and sanitation related diseases like diarrhea each year (Water Aid, 2009). Poor sanitation led to Indian GDP reduction by 6.4 per cent equating to a loss of $53.8 billion in 2006 (World Bank, 2008; WSP, 2010). Improved household sanitation, when sustainable and equitable, not only benefits health and economy, but also provides environmental cleanliness, safety, privacy and dignity for all (Water Aid, 2009).

While the statistics are disheartening and the implications unambiguous, there is hope of universal household sanitation soon emerging in India. One of national goals under developmental projects in India is to provide and facilitate the adequate water supply for drinking and cooking on a sustainable basis to every individual household. Only access to drinking water is not enough to reduce health problems unless the quality of drinking water is ensured. Bacteriological contamination in drinking water is the major source for child mortality. Therefore, it is equally important to ensure portability and reliability of drinking water quality standard both at the production and consumption level.

Due to India's household sanitation shortcoming, the Government of India (GoI) launched the Central Rural Sanitation Programme (CRSP) in 1986 to lead water, sanitation and health progress in rural India. The CRSP adopted the Total Sanitation Campaign (TSC) in 1999 to accelerate sanitation progress including coverage of household latrines and other sanitation components (Water Aid, 2008; GoI, 2010). The TSC aims to be a demand-driven, people-centred approach characterised by Information, Education and Communication (IEC) for social mobilisation and by ₹ 2200 per household hardware subsidy in plains to motivate toilet construction and use in below poverty line (BPL) household (GoI, 2010). The subsidy is not available to above poverty line (APL) households. While this conventional TSC approach is most common throughout India, it is not ubiquitously applied. As part of the TSC, an alternative approach characterised by absence of household hardware subsidy and by participatory rural appraisal triggering tools for evocation of disgust, shame and humiliation to motivate behaviour change is now applied (Water Aid, 2008). Facilitating organisation have begun utilising the TSC's alternative approach, known as Community Led Water Safety Environment & Sanitation, to varying degrees in rural areas of at least 9 of 28 Indian states including Gujarat, Haryana, Himachal Pradesh, Madhya Pradesh, Maharashtra, Meghalaya, Orissa, Rajasthan and Uttarakhand.

In the same context, India Developed Marketplace (IDM) which is an initiative of the World Bank in promoting innovative ideas in the field of development through a competitive process on an annual basis. The study presented is an awardees' of the same initiative, where under India's National Rural Drinking Water Quality Monitoring & Surveillance Program (NRDWQMSP), Community Led Action for Water Safety & Environment Sanitation (CLAWSES) has been developed and implemented as an award scheme in one block in India. It has the support of the World Bank. Self-motivated panchayati raj institutions (PRIs) members were more effective in facilitating CLAWSES exercises than the non-governmental organisation (NGO) staffs. It is amenable to be replicated at other places also.

The Choice of Nainital

The idea of CLAWSES was applied within the context of NRDWQMSP on a pilot basis in Nainital District in Uttarakhand. The initiative was found to be innovative and selected for implementation and finally, it was decided to implement the project in 'Dhari' block of Nainital District (Figure 11.1). Another reason for the choice was that the area was close to the key resource centre (KRC) situated at Administrative Training Institute (ATI), Nainital, and facilitated easy monitoring of the project. An action plan was prepared by the District Project Management Unit (DPMU), Bhimtal, with the assistance of partner institute of KRC to implement the project during November 2008 to September 2009. A reward ceremony was organised on February 20, 2010, and Mr S. Raju, Commissioner and Director of ATI, Nainital, distributed the rewards to the four winning *gram panchayats* (GPs).

Activities Undertaken

As part of the project implementation, several activities were conducted between November 2008 and 2009 such as kick-off and preparatory meetings were held to create a shared common understanding about the project between KRC, Centre for Good Governance (CGG), Uttarakhand Academy of Administration (UAoA), PMU, Swajal, Uttarakhand and DPMU, Bhimtal, in November 2008. It led to the preparation of an action plan by DPMU, Bhimtal, with the assistance of KRC for implementation in Dhari block of Nainital District and its approval by PMU, Swajal. A draft of the block reward scheme was prepared and selection of facilitators done by DPMU for facilitating the process in the villages of Dhari block. Simultaneously, arrangement of hydrogen sulphide (H_2S) vials, sanitary survey formats was done and a four-day training programme on CLAWSES methodology was facilitated by

KRC at Dhanachuli in Dhari Block to train the selected facilitators. These trained facilitators then committed to start working initially in 19 GPs. A first round of follow-ups was then done by making phone calls with team leaders and community leaders and PRI members, and making visits to villages and interacting with communities.

A number of activities were initiated by communities. These included: sanitary surveys and H2S vial tests carried out by the community members in 70 water sources of 16 GPs; communities contributing labour for cleaning of some 5 traditional water sources in two GPs (namely, Dhanachuli and Sunderkhal); community members started collecting contribution of ₹ 10/HH/month for maintenance of water sources in 2 GPs; community banning of washing of clothes near the *naula* (traditional water source) in Khad hamlet of Dhanachuli GP; community members initiating formation of vigilance committees for ensuring cleanliness at water sources and stopping open defecation and construction of temporary toilets in 2 GPs; community putting signboards to stop open defecation and punishment of ₹ 50 for violators in 2 GPs; rally by some 300 children in Dhanachuli GP shouting slogans and women discussing menstrual hygiene in one GP, etc.

The series of activities were followed by a meeting between DPMU and KRC team to review the progress of preparations for launch of the reward scheme. Finally, block reward scheme was launced at Dhari block office by Commissioner, Kumaon, in the presence of *gram pradhans*, other BDC members, district/block officials and media representatives. This triggered CLAWSES and second round of follow-ups by KRC team along with DPMU staff. The GPs covered included Budibana, Sasbani, Sunderkhal, Managher, Aksoda, Dhanachuli, Kokilbana, etc. An elaborate meeting between DPMU and KRC was conducted to discuss as to how to complete the IDM project in view of the verbal instructions of PMU to stop working on the project as water quality programme being transferred to Jal Nigam. Consequently, a revised action plan was prepared by DPMU with the assistance of KRC for timely completion of the IDM project.

The KRC team then initiated the mentoring and third round of follow-ups. The GPs covered included Saliakot Talla and Malli, Dinni Talli and Malli, Managher, Kaul, Gajjar, Sunkia, Gunia Lake, Devnagar, Babiyad, Dudli, Jalnilpahari, Mahtoliagaon, Majuli, Sakdina, Saran, Aghria, Latfoda, Buranshi, Harinagar, Sunderkhal, Dhanachuli and Panyali, etc. Many activities were carried out by GPs including sanitary surveys, H2S vial tests, improvement/cleaning/protection of water sources, cleaning of drains, construction of toilets and stopping open defecation, upgradation of toilets, etc. Natural leaders from some GPs such as Dhanachuli, Sunderkhal, Managher helped the communities of other villages to understand the importance and process of community-led action for water safety and environment Sanitation of water sources.

Table 11.1 shows the GP-wise information of water sources with their risk levels as per sanitary surveys conducted during February-July 2009 by respective communities in GPs of Dhari blocks, Uttarakhand, India.

Table 11.1

GP-wise Information of Water Resources

S.No.	*Gram Panchayat*	*No. of Water Sources*	*Risk of Water Sources based on Sanitary Survey*				*No. of Water Samples found Unsafe for Drinking in 1st Time Testing*	*No. of Water Samples found Unsafe for Drinking in 2st Time Testing*
			Very High	*High*	*Medium*	*Low*		
			No.	No.	No.	No.		
1.	Dinni Malli	10	-	1	4	5	7	1
2.	Gunia Lake	4	-	-	1	3	3	-
3.	Majuli	6	-	1	3	2	4	-
4.	Saliakot Malli	9	-	1	5	3	2	-
5.	Jalnilpahari	6	-	-	2	4	4	-
6.	Silalake	12	-	2	8	2	5	1
7.	Devnagar	45	-	8	24	13	32	1
8.	Sunkia	8	-	1	3	4	5	1
9.	Parvada	10	0	0	0	-	6	2

.contd...

...contd...

			No.	No.	No.	No.		
10.	Kaul	9	-	-	-	9	5	-
11.	Sakdina	14	-	2	3	9	6	-
12.	Saliakot Talla	5	-	1	2	2	4	-
13.	Maghera	10	-	3	3	4	2	3
14.	Mahtoliagaon	13	-	1	9	3	9	-
15.	Latfoda	5	-	-	2	3	2	-
16.	Buranshi	5	-	-	2	3	2	-
17.	Dudli	8	-	-	6	2	3	-
18.	Babiyad	11	-	2	4	5	9	-
19.	Aghria	9	-	1	3	5	5	-
20.	Panyali	9	-	-	3	6	7	-
21.	Harinagar Aksoda	10	-	1	4	5	4	-
22.	Budibana	17	2	-	3	12	5	2
23.	Dhanachuli	16	-	6	10	-	1	1
24.	Sasbani	6	-	5	-	1	1	1
25.	Managher	15	4	1	10	-	13	2
26.	Kokilbana	4	-	1	2	1	1	-
27.	Sunderkhal	21	-	3	14	4	4	1
28.	Gajjar	5	1	1	1	2	2	-
29.	Chakhuta	9	-	2	2	5	5	-
30.	Aksoda	17	-	1	5	11	14	-
31.	Dinitalla	5	-	-	-	-	3	-
32.	Sarna	8	-	-	6	2	5	=
	Total	341	7	45	144	130	180	16

Finally, DPMU initiated the process of inviting GPs to apply for the reward and evaluation committees were formed. The committees comprised of members from the block, Jal Nigam, Jal Sansthan, Dhari block Kshetra *panchayat* and representative of local media; approval of the committee was done by the Chief Development Officer, Nainital, and a one-day orientation of evaluation committee was conducted. It was unique in the sense that all the GP/VWSC representatives were invited (22 members from 17 GPs attended) and they were asked to plead their cases as to how they deserved to be given the rewards. Those GPs which could not apply during the stipulated deadline for the same were given the last chance to apply. Each and every case was discussed and 15 out of 28 GPs were shortlisted to be physically verified for evaluation and the remaining were *prima facie* rejected based on the criteria laid down for the purpose.

The shortlisted GPs were evaluated by the evaluation team in December 2009. Marks obtained by the GPs are given in Table 11.2.

Table 11.2

Marks Obtained by the GPs

S.N.	*Name of shortlisted GP*	*Marks obtained against Maximum Marks of 100*
1.	Sunderkhal	86
2.	Devnagar	79
3.	Sasbani	75
4.	Managher	72
5.	Gunia Lake	58
6.	Dhanachuli	54
7.	Mahtoliagaon	48
8.	Kokilbana	44
9.	Budibana	45
10.	Parvada	44
11.	Majuli	41
12.	Maghera	37
13.	Dudli	30
14.	Dinni Talli	36
15.	Babiyad	44

Finally, the decision of the DPMU about the names of the four winning GPs was communicated to the KRC on January 7, 2010. The reward distribution ceremony was organised by the DPMU at Dhari block in which the four winning GPs were felicitated and given the bank drafts of rewards by the Commissioner, Kumaon.

Results and Discussions

The following are the key results achieved in the project:

- Communities in most GPs (32 out of 35 in the block) assessed the risks associated with the drinking water sources by carrying out sanitary survey of all (341 water sources in 32 GPs) the sources that were being used by them for drinking and cooking in their respective areas.
- Communities also tested all the 341 water sources in the 32 GPs for their bacteriological contamination using H2S vials. These tests were carried out between February and July 2009.
- In 14 out of 32 GPs, communities undertook H2S vial tests of water sources again during monsoon i.e., during July-August 2009. Some 7 GPs undertook water testing third time also.
- A total of 10 GPs (Sunder, Khalk, Dhanachuli, Devnagar, Kokilbana, Buranshi, Managher, Sasbani, Gajjar, Parvada and Chaukhuta) reportedly have become open defecation free without any external assistance. Some of them started with simple low-cost toilet options and gradually moved to more durable *pucca* structures. Fine to defaulters was declared by the communities in order to stop open defecation in villages and in schools. In cases to combat water scarcity, children started the practice of carrying water in bottle from home to use for toilet in schools.
- Communities in almost all the 32 GPs undertook one or more protective and remedial measures based on the results of sanitary survey and H2S vial testing of water sources. These included: repairing the broken platform of hand pumps, boiling of drinking water, keeping water pots covered, use of chlorine tablets for treating water at household level, chlorination at community level by putting bleaching powder in *naula* (traditional water source) and clear water reservoir (CWR), cleaning and protecting of water sources and CWRs, contacting employees of Jal

Sansthan to get bleaching powder and requesting them to unlock the CWR for cleaning purpose, changing the defecation practice, start entering in *naula* without shoes/slipper while filling water containers and improved cleanliness in households.

- A number of natural leaders emerged during the process that not only led the actions in their own villages but also helped people in other villages to undertake water quality monitoring and surveillance in their respective villages.

Highlights of the Initiative

- Communities are able to conduct sanitary survey and bacteriological testing of water source (using the H2S vials), if facilitated properly.
- Launch of a reward scheme creates a spirit of competition among members of PRIs to institutionalise water quality surveillance in villages.
- CLAWSES methodology is an empowering process that helps communities in tackling the issue of open defecation as well as water quality.
- Awareness level of villagers increased regarding waterborne diseases and need to protect water sources.
- In CLAWSES intervention villages, money issue becomes secondary and health and dignity primary after communities realised the linkage between practices related to defecation, water management and diseases.
- Natural leaders emerge during the process of CLAWSES exercise that undertook responsibility first in their village and later in the neighboring hamlets and GPs. Women and young girls (students) were found to be more active in mobilising community.
- Those who stood in opposition to community's own initiative to water management later joined the bus when peer pressure mounted through work done by larger community, natural leaders and PRIs.

- The sense of solidarity increased in CLAWSES intervention villages, which is otherwise diminishing in villages.
- The facilitating role and commitment of implementing agency at the block/district level is critical in the spread of the CLAWSES across various GPs. The lukewarm response of PMU/DPMU during certain periods of implementation created hindrance and slowed down the pace of implementation. Frequent transfers of decision-makers in DPMU also affected the progress and resulted into delays.
- Follow-ups (in consultation with and demand of the natural leaders and GP) are critical in facilitating institutional mechanism, remedial and preventive measures at the village level as well as to ensure sustainability of outcomes achieved.
- Self-motivated PRI members were more effective in facilitating CLAWS exercises rather than a few NGO staff who showed reluctant attitude throughout the project period, mainly because the DPMU could not develop an incentive mechanism for their work.
- The results of CLAWSES approach and methodology of rewarding outcomes at the community level are encouraging and can be replicated in other blocks of Uttarakhand and India. Since CLAWSES approach is based on community initiative, community is managing whole expenditure for testing, cleanliness and cost of bleaching powder for disinfection of bacteriological contamination. Presently all GPs have formed local committee and every family is contributing ₹ 5.00 each month. Therefore, community itself insuring sustainability.

Conclusion

The study highlighted how self-motivated PRI members were more effective in facilitating CLAWSES exercises than the NGO staffs. It also brought to fore that the communities are able to conduct sanitary survey and bacteriological testing of water source (using the H2S vial), if facilitated properly. Also if communities are made to realise the linkages between practices related to defecation, water management and disease, then the health and dignity issues become primary and money becomes secondary. Through this process, community is empowered to tackle the issues of open defecation and water quality. The sense of solidarity and social cohesiveness increased in sample villages which are otherwise diminishing in today's societies.

The results of CLAWSES approach and methodology of rewarding outcomes at the community level are encouraging and can be replicated in other blocks of Uttarakhand and India. The approach is based on community initiative, which manages the whole expenditure for testing, cleanliness and cost of bleaching powder for disinfection of bacteriological contamination. Such efforts if initiated in the right way can forge the way for solving much of India's water and sanitation problems.

References

Government of India (2010). *Guidelines Central Rural Sanitation Programme Total Sanitation Campaign*. June.

WaterAid (2008). *Feeling the Pulse: A Study of the Total Sanitation Campaign in Five States*. New Delhi: WaterAid India.

———. (2009). *India-Country Information*. Retrieved December 14, 2010 from *http://www.wateraid.org/documents/india_a4.pdf*

———. (2010a). *Statistics*. Retrieved December 14, 2010 from *http://www.Wateraid.org/international./what_we_do/statistics/default.asp*

———. (2010b). *A Silent Transformation: Total Sanitation Campaign in Haryana*. New Delhi: WaterAid India.

World Health Organization and UNICEF (2010). *Progress on Sanitation and Drinking Water: Update 2010*. WHO/UNICEF Joint Monitoring Programme for Water Supply and Sanitation. Geneva: World Health Organization.

World Bank (2008). *Improved Sanitation Facilities*. Retrieved December 28, 2010 from http://data.worldbank.org/indicator/SH.STA.ACSN/countries

Water and Sanitation Programme (2010). *What were the Total Economic Impacts due to Inadequate Sanitation in India?* New Delhi: World Bank.

Acknowledgements: The authors would like to extend thanks to World Bank India Office; Department of Drinking Water Supply, Government of India; District Project Management Unit, Bhimtal; Knowledge Links Private Limited; Mr S. Raju, IAS, Director, UAoA, and Commissioner Kumaon, Nainital; Mr V.K. Mishra, Deputy Director and Coordinator KRC, UAoA, Nainital and PMU, Swajal Project, Dehradun.

12

MANISH KUMAR and VIJAY MITTAL

Sanitation in Uttar Pradesh

Current Status and Strategic Choices

Introduction

Uttar Pradesh (UP) is one of the largest states of India. The total area of the state is 243,290 sq km with a population of approximately 201 million (2010). There are only five countries in the world with more population than UP, highlighting the significance of this region with respect to global human development indicators. The northern part of the state is mountainous, middle alluvial and flatland, and the southern part a dry plateau. The central part is the largest part of the state and also the most populated, with river Ganges flowing right across.

The state domestic product (2005) of UP was ₹ 242,000 crores, making it the second largest economy within India contributing nearly 8 per cent to India's GDP. The state has been growing at nearly 6.5 per cent per annum. However, the per capita income (2005) was ₹ 10,637, which is less than the national average of ₹ 20,734. The contribution of primary, secondary and tertiary sector to the economy is 43 per cent, 20 per cent and 37 per cent, respectively.

The objective of the present study is to analyse the impact of non-achievement of Millenium Development Goal (MDG) on the economy of the state. Additionally, the study suggests a road map for UP to accelerate its progress towards MDG on sanitation.

Sanitation Status in Uttar Pradesh and Policy Analysis

The population of UP in 2006 was 185 million. It is the most populous state in India; 16.5 per cent of country's population is concentrated here. With almost 125 million population of the state practising open defecation (Figure 12.2), UP will have significant impact on India's ability to achieve MDG in sanitation by 2015. One out of every five persons in the country who defecate in the open live in UP.

Figure 12.1

Population Projections for Uttar Pradesh (Million)

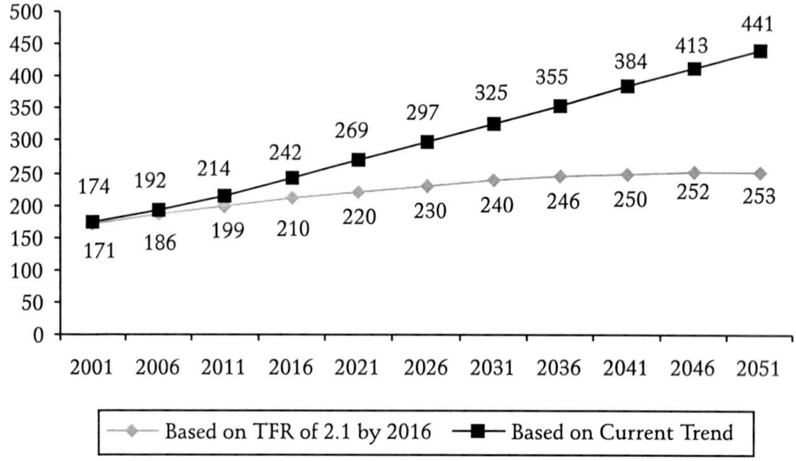

Source: Population Policy of Uttar Pradesh, Department of Health and Family Welfare, 2000.

Figure 12.2

State Level Disaggregation of Number of Open Defecators

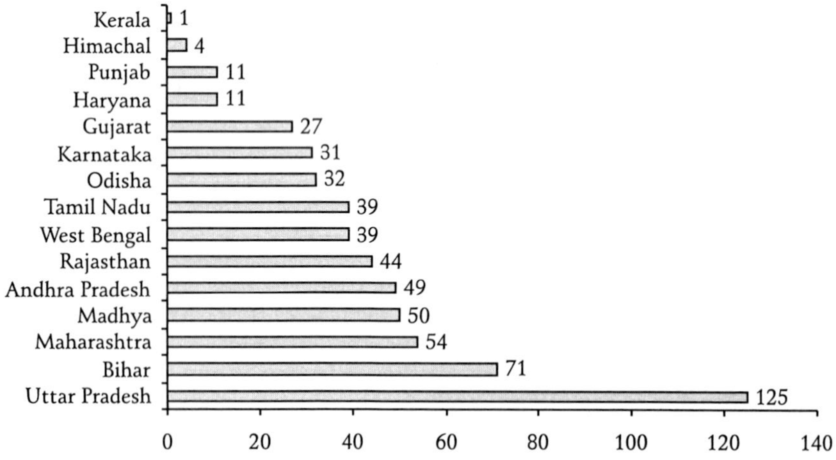

Source: Computed by Authors.

As evident from Figure 12.2, UP tops the states in terms of open defecators; Bihar is a distant second.

Analysis of sanitation data using Joint Monitoring Programme (JMP) methodology shows that MDG target for sanitation in the case of UP stands at 56.5 per cent. In other words, at least 56.5 per cent population in UP should be using improved toilet by 2015 if the state were to achieve MDG target in sanitation. In 2006, toilet use was only 32.4 per cent and use of improved sanitation facility was even lower at 19.6 per cent. These figures have been corraborated by the concurrent monitoring of sanitation supported by the United Nations International Children's Emergency Fund (UNICEF) in UP. A random sample of 6,443 households showed that only 32 per cent use toilets, 8 per cent toilets are defunct and 60 per cent households do not have toilets of any kind. Importantly, all households members were found using toilets only in case of 23 per cent of the sampled households. This sample size is small for a state like UP but the findings are generally consistent with data from previous National Family Health Survey (NFHS) and District Level Household Survey (DLHS).

Figure 12.3 is prepared using JMP methodology. It shows the trend of improved sanitation (rural plus urban) since 1990. The estimates for various years have been derived using a regression line. This methodology precludes overstatement of performance by averaging findings from different surveys, recognising that these surveys may be measuring toilet use differently. A sudden increase shown by a single survey during a particular year is, therefore, not over-emphasised.

Data from various national surveys since 1990 have been taken into account for computation. There has been a mere 6.9 percentage point increase in improved sanitation over 15 years.

Figure 12.3

Access to Improved Sanitation Coverage: Total UP

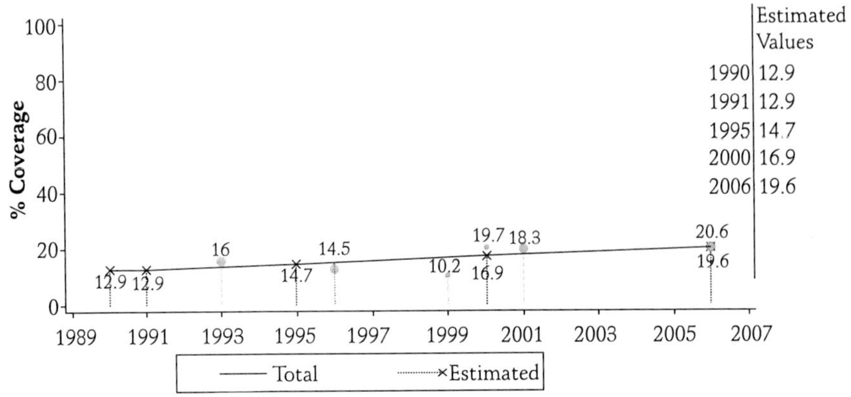

Note: 1993: NSS49r, 1996: NSS52r, 1999: NFHS2, 2000: MICS, 2001: Census, 2006: NFHS3.

The percentage of open defecators has declined from 78.0 per cent to 67.6 per cent—a decline of 10.4 percentage points, as seen in Figure 12.4. Evidently, a significant number of new toilet users have adopted unimproved sanitation technology which is unsafe from health perspective. Thus, the society is not able to benefit fully from adopting toilet use behaviour.

Figure 12.4

Coverage under Open Defecation: Total UP

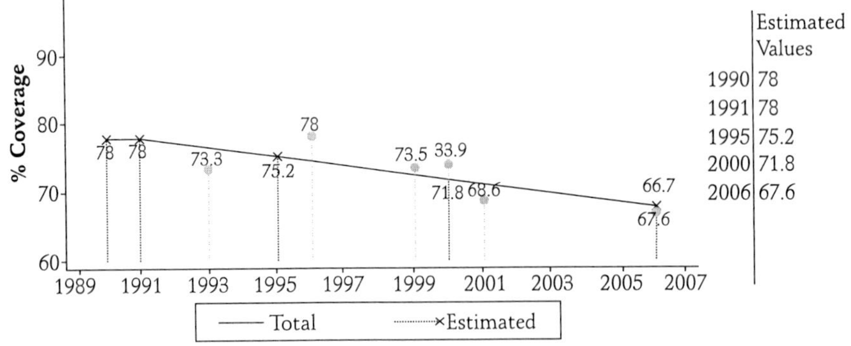

Note: Same as Figure 12.3.

The rural population of the state is 147 million, with only 11.8 per cent having access to improved sanitation (Figure 12.5). Although the number of improved toilet users has almost tripled since 1990, current level of efforts has proven insufficient given the huge initial gap and rapid population growth.

Figure 12.5

Access to Improved Sanitation Coverage: Rural UP

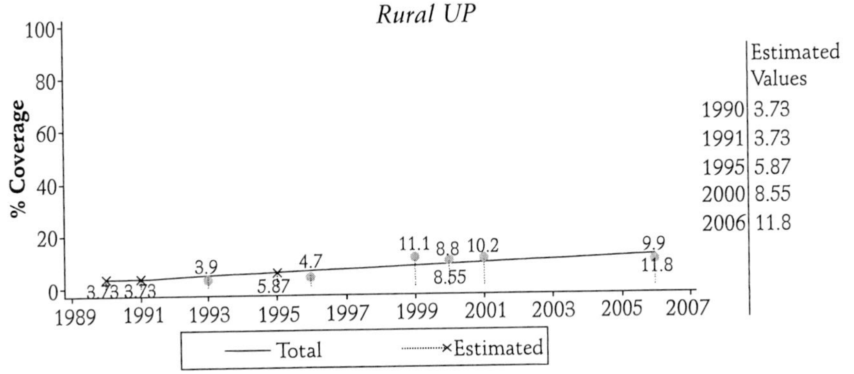

Note: Same as Figure 12.3.

The urban population is about 38 million and improved sanitation amongst the urban population stands at 59.8 per cent. Although vastly better than rural sanitation, there has been only a 4 percentage point increase in improved sanitation in urban areas in 16 years, indicating almost insignificant improvement over time.

Figure 12.6

Access to Improved Sanitation Coverage: Urban UP

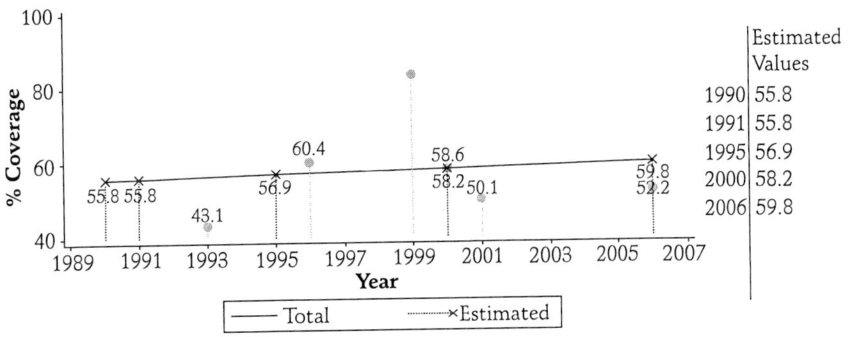

Note: Same as Figure 12.3.

Achieving MDG Targets

If MDG is to be achieved, TSC must be accelerated in selected districts.

There are 71 districts in UP. At the current rate, improved sanitation will stand at 24 per cent in 2015 and improved plus unimproved sanitation will be 39 per cent. In either case, the achievement will be short of MDG target of 56.5 per cent. The status in 2015 is tabulated below:

1.	Projected population of UP in 2015	220 million
2.	MDG Target (56.5%)	124.5 million
3.	MDG achievable at current rate (24%)	53 million
4.	Unimproved toilet users (39% - 24% = 15%)	33 millon
5.	Open defecators in 2015	135 million
6.	MDG target shortfall	71.5 million

This target shortfall can be covered in two ways:

- Converting non-improved toilet users into improved toilet users. The maximum increase possible is 33 million, which is less than target shortfall at status 6 above.
- Converting open defecators to improved toilet users. This is the primary focus of Total Sanitation Campaign (TSC) and is, therefore, the policy instrument on which India is relying to achieve its MDG goal.

If MDG shortfall is to be covered in the remaining 5 years, a pertinent question is what are the regions of the state in which the government should focus its resources to maximise outcome. DLHS 3 provides data for 2008 of pecentage of households in each district having access to sanitation (construed as toilet use in JMP methodology). Analysis of DLHS 3 data, alongwith population data from Census 2001 (population projected to 2008), shows that nearly 11 per cent of the open defecators are concentrated in four districts of UP, each district having nearly 3.40 million open defecators or more. If 100 per cent improved toilet use were to be achieved in these four districts, the gap in MDG for sanitation, as on 2015, would have to be reduced by more than 15 million.

These groups of four districts (Group 1) are listed in Table 12.1 in decreasing order of open defecators. Allahabad has the highest concentration of open defecators followed by Azamgarh, Jaunpur and Sitapur. The first three districts are contiguous, indicating the concentration of toilet use problem in a geographical locality.

Table 12.1

Districts in Group 1: Uttar Pradesh

District	*Projected Population of Open Defecators (2007)*
Allahabad	4,084,668
Azamgarh	3,741,821
Jaunpur	3,651,550
Sitapur	3,429,279

Note: Figures based on the assumption that post 2001 census, population is increasing in the districts by 2 per cent per year.

DLHS 3 data also indicate that there are 18 districts in UP with 2.40 million open defecators or more. Together, these districts account for 37 per cent open defecators in UP. If 100 per cent improved toilet use were to be achieved in these 18 districts, the gap in MDG as on 2015 would have to be reduced by further 49.5 million. These 18 districts (Group 2) are listed in Table 12.2 in decreasing order of open defecators.

It is interesting to note that districts with lowest per cent of households having access to toilet facility may not figure in the list given in Table 12.2. For example, the following districts have less than (or equal to) 10 per cent households with access to sanitation facility as per DLHS 3: Ambedkar Nagar (9.7%), Gonda (9.8%), Pratapgarh (10%), Shrawasti (8.4%) and Siddarthnagar (6.9%). But only two of these—Gonda and Pratapgarh—figure in Table 12.2 because the absolute numbers of open defecators in the rest of these districts are not very high.

While the ranking proposed above is deliberately designed around absolute number of people that could be impacted, given that MDGs are set against population to be covered (and, therefore, the number of people to be covered is critical to its achievement), there are concerns about equity in following a strategy-based solely upon absolute numbers. Inclusive growth principles may be compromised if ranking of districts as given above is to solely determine strategy. For example, it is not clear, given the data constraints, that how many scheduled castes (SCs) and scheduled tribes (STs) will be benefited if top four districts (Table 12.1) in terms of absolute numbers were to be made open-defecation free. Therefore, a balanced strategy is required to be evolved in which emphasis on absolute numbers to be covered is supported by strong monitoring mechanisms to track coverage of SCs and STs in these districts.

Table 12.2

Districts in Group 2: Uttar Pradesh

Name of District	*Projected Population of Open Defecators (2007)*
Gorakhpur	3,217,968
Hardoi	3,194,743
Sultanpur	3,097,598
Kheri	3,002,027
Ghazipur	2,908,639
Rae Bareli	2,846,420
Kushinagar	2,817,120
Gonda	2,809,449
Agra	2,781,767
Pratapgarh	2,764,099
Ballia	2,625,415
Etah	2,574,837
Deoria	2,561,347
Budaun	2,561,243
Barabanki	2,522,946
Unnao	2,472,429
Aligarh	2,417,982
Bahraich	2,352,095

Note: Figures based on the assumption that post 2001 census, population is increasing in the districts by 2 per cent per year.

Children below 14 years constitute nearly 37.1 per cent of UP's population. This is above the all India average of 30.6 per cent. Thus, the chances of influencing toilet use through awareness building in schools are much higher in the case of UP compared to other states.

The population of 65 years and above constitute 4.3 per cent of UP's population which is below the national average of 5.2 per cent. Although the chance of changing behaviour of this age group is the least, this is, however, not a problem for UP given their small number. Thus, UP population age profile is favourable for undertaking large-scale awareness generation campaign in respect to sanitation, with potential for multi-generational impact.

Policy Variables impacting Toilet Use

Insights on policy variables that impact toilet use can be helpful in developing strategies for MDG on sanitation. Awareness of sanitation practices, surplus household income, government investments and access to water are important determinants that influence toilet use decisions. Awareness of health and sanitation linkages motivates households on toilet use. Surplus income is essential to undertake construction of toilets. Government investments through TSC are made with the expectation that it will speed up toilet use. Lack of access to water is an often cited reason that why open defecation is preferred to constructing toilet in the household.

DLHS 3 district-level cross-sectional data of 70 districts in UP were analysed for that above variables impacting toilet use at household levels. A linear regression model with multiple variables was used. The dependent variable was toilet use, while independent variables included proxies for sanitation awareness, poverty, TSC expenditure and access to drinking water. The expectation in respect of sanitation practice awareness was that a large proportion of literate population in a district would be positively associated with toilet use. The proxy used was male/female literacy; greater levels of literacy in a district should imply greater levels of awareness about toilet use.

In respect to poverty, the expectation was that poverty level would be negatively associated with toilet use; districts with high below poverty line (BPL) card holders should show low toilet use.

In respect to government investments, the expectation was that districts that have made high-level TSC expenditure per capita over years would show positive association with toilet use. The cumulative central share of TSC fund allocated to districts up to March 2007 was taken for analysis. District population from Census 2001 data was used for computing per capita expenditure. In respect to water, the expectation was that access to piped water supply would be positively associated with toilet use.

The findings of the analysis is presented in Table 12.3. Five regression models were run using various combination of independent variables.

Table 12.3

Independent Variables Impacting Toilet Use

Dependent Variable: "Households have Access to Toilet Facility (%)"						
Independent Variables	*Model 1*	*Model 2*	*Model 3*	*Model 4*	*Model 5*	*Remarks*
Per cent literate Female Population (Age 7+)	0.85 (0.21)***		0.48 (0.17)***	0.42 (0.16)***	0.25 (0.18)*	
Percent literate Male Population (Age 7+)		0.64 (0.29)**				
Households have BPL Card (%)			-0.85 (0.11)***	- 0.74 (0.11)***	-0.63 (0.14)***	
Dist cumulative TSC Expenditure per capita				-0.13 (0.04)***	-0.12 (0.04)***	
Households use piped Drinking Water (%)					0.50 (0.18)***	
Constant	-17.68	-22.32	25.30	27.94	29.86	
R square	0.19	0.08	0.56	0.61	0.65	
Adjusted R square	0.18	0.07	0.54	0.59		
N	70	70	70	70	70	

Note: *** significant at 1 per cent level; ** significant at 5 per cent level ; * significant at 10 per cent level.

Female Literacy and Toilet Use

Literacy appears to have a positive association with toilet use. Although literacy is a general term, Communication for Development (C4D)—critical for sanitation behaviour change—is expected to show higher impact in a literate society. Model 1 shows that female literacy is positively and significantly associated with toilet use. Model 2 finds that male literacy is also positively associated with toilet use, although the level of significance is lower (at 5% level compared to 1% for female) and its explanatory power is low too (lower adjusted R square). As male and female literacy are found to be strongly correlated (92%), Model 3 takes only female literacy because of its superior explanatory power. In both Models 3 and 4, which take into account additional independent variables, female literacy is found significant at 1 per cent level. As per Model 4, one per cent increase in female literacy would result in increase in toilet use by 0.42 per cent. In Model 5, female literacy remains positively associated but is significant at 10 per cent level. In this modelling, 1 per cent increase in female literacy of a district is found associated with 0.25 per cent increase in toilet use of the corresponding district.

A scatter plot of female literacy and toilet use covering all 70 districts is given in Figure 12.7. As observed from the slope of the line, there is a positive relation between the two. The overall inference is that female literacy is positively associated with toilet use.

Figure 12.7

Access to Sanitation versus Female Literacy

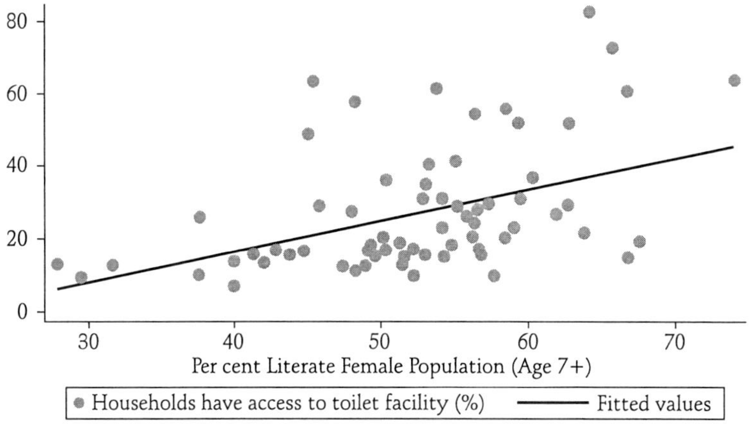

The policy insight is that awareness campaign for toilet use should be taken up first in those *panchayats* where women literacy is high. The chances of success in such *panchayats* are the highest.

Impacts of TSC Expenditure on Toilet Use

The regression analysis shows that per capita TSC expenditure in a district is negatively associated with toilet use decision. The coefficient is significant at 1 per cent level. This may be pointing towards a serious flaw in the assumption that higher level of toilet use is achievable by increasing TSC expenditure. In fact, the contrary appears to be the case. An increase in per capita expenditure by 1 per cent in a district is associated with 0.12 per cent lower toilet use in district, holding all other parameters constant.

The analysis in this paper in respect to expenditure requires to be substantiated by more studies. A rigorous time series analysis would reveal much more than cross-sectional analysis undertaken for this paper. Further, TSC is for rural areas while toilet use data is for both urban and rural. Hence, TSC expenditure does not reflect fully government investment in sanitation and, although taken in this analysis as a proxy, is not the most appropriate independent variable for the analysis. Given data limitations, this was the best that could be attempted at this stage. Finally, IEC expenditure requires to be regressed to understand if there is positive relation between current IEC activities and toilet use behaviour.

Scatter plot of per capita TSC fund distribution in districts against households having access to toilet facility is given in Figure 12.8.

Figure 12.8

Per Capita TSC Expenditure for Districts

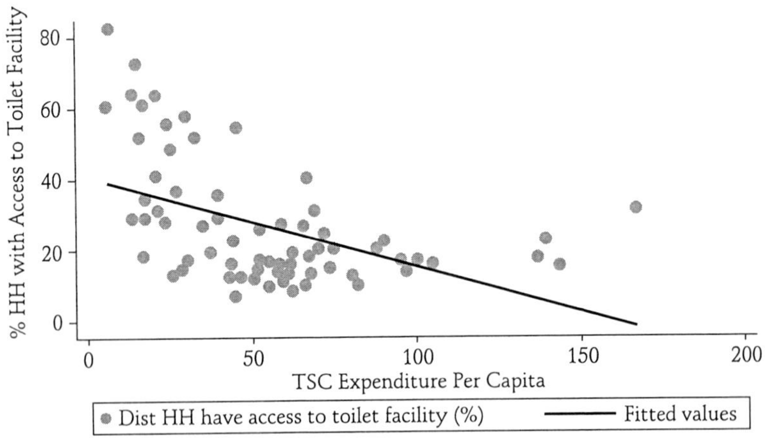

Importance of Water Accessibility

Piped water to households appears to be positively and significantly associated with toilet use in a district. Model 5 has higher adjusted R square compared to Model 4, indicating that water availability adds explanatory power to the modelling of toilet use behaviour. The coefficient is significant at 1 per cent level. One percent increase in household piped drinking water in a district is associated with 0.50 per cent increase in toilet use behaviour.

An insight from the above analysis is that water availability is important for promoting toilet use behaviour. Alternatively, innovative technological adaptations are essential for sanitation facilities that need less water.

Figure 12.9

Access to Toilet Facility versus Water Supply

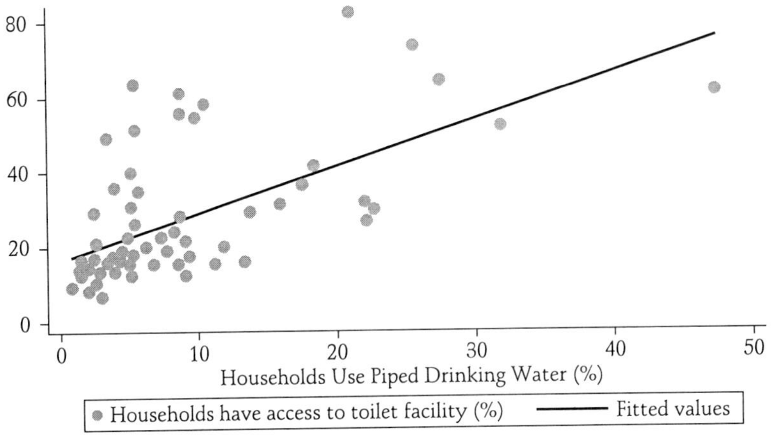

Poverty and Toilet Use Behaviour

Poverty card (Figure 12.10) is negatively associated with toilet use behaviour in a district and is statistically significant at 1 per cent level. The addition of poverty proxy as an independent variable added explanatory power to the model. As evident from Table 12.3, an increase of BPL card holders by 1 per cent in a district results in a decrease of toilet use behaviour by 0.63 per cent.

Income Profiles in Uttar Pradesh

It appears that awareness relating to health and sanitation linkage on its own may not induce a household to take up toilet construction and use. The households may simply lack income to take up construction.

The income profile of households in rural UP was analysed to further understand how strongly poverty constrains choices, including choice to construct and use toilets. The household expenditure data from a survey commissioned by Government of India (GoI) provides good insight.[1] The data is presented in Tables 12.4 and 12.5.

Figure 12.10

Access to Toilet Facility versus Poverty

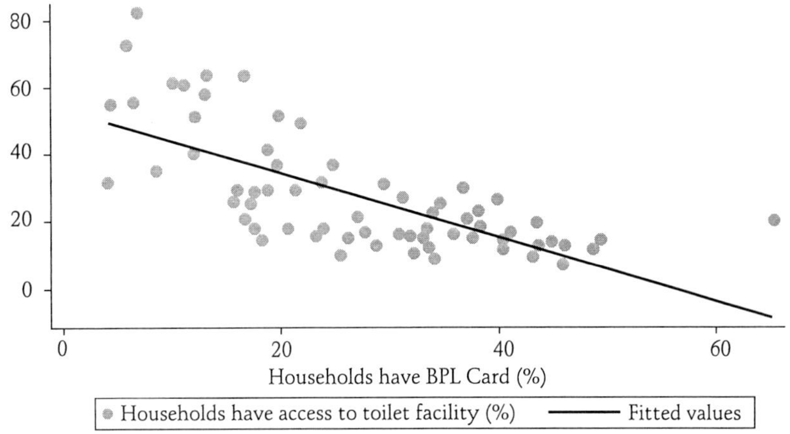

It is observed that 61 per cent households of rural UP have a monthly disposable income of less than ₹ 1,691. The average income in this category is ₹ 873. About 20 per cent have income less than ₹ 3,000 but greater than ₹ 1,691. The average in this category is ₹ 2,185 per month. Although not directly imputed by the survey, open defecators in rural UP are likely to be concentrated within these two income bands.

1. Survey of UP accessible at: *http://www.csc-india.org/AboutCSCProject/ACNielsonReports/tabid/704/language/en-GB/Default.aspx*.

Table 12.4

Distribution of Households by Mean Monthly Expenditure

Expenditure Patterns of Surveyed Households		*% of Households*
Total Monthly Expenditure (₹)	<1691	60.9
	>1691 - < 3001	20.4
	>3000 - <5001	9.8
	>5000 - <7500	4.0
	>7500	4.9
Total		100.0

Table 12.5 gives the expenditure behaviour of households in different income bands. The lowest income band appears to live in extreme indigence. Food and household items consume 65 per cent of their income, 11 per cent is on medical issues and 8 per cent on transportation. A toilet constructed at ₹ 2,500 is equal to nearly three months expenses for these families.

Table 12.5

Mean Expenditure of Households Surveyed in Uttar Pradesh

	Total Monthly Expenditure (₹)					*Total*
	<1691	*>1691 - < 3001*	*>3000 - <5001*	*>5000 - <7500*	*>7500*	
Transport—Bus pass	1.26	2.01	7.40	8.11	34.83	3.92
Other transport cost	69.51	177.03	360.28	576.51	1,662.87	217.81
Household goods (milk, grocery, other food items, electricity, clothes, etc.)	568.42	1,329.95	1,936.50	2,551.39	8,534.60	1,324.71
House rent	4.62	4.68	13.58	7.38	70.03	8.80
Education—School	39.08	80.79	153.84	291.11	5,403.86	329.58
Education—College	12.35	51.68	156.45	298.47	626.62	75.89
Education—Tuition	7.82	23.26	62.21	85.67	142.92	26.01
Medical expense—For common illnesses (Fever/Cough/Diarrhoea, etc.)	90.98	192.36	290.63	507.92	1,071.51	195.68

contd...

...contd...

Medical expense—Diagnosis for pregnancy, infant issues, child diseases, etc.	2.51	10.36	36.72	126.65	873.43	54.79
Medical expense—Advanced treatment for TB/malaria/Pneumonia, etc.	1.82	9.17	14.76	15.02	198.25	14.66
Medical expense—Advance treatment for heart diseases/cancer, etc.	0.36	1.56	11.09	0.80	76.61	5.38
Medical expenses—On diagnostic tests (blood reports, X-rays, ultra sound, etc.)	0.78	5.13	9.63	27.92	355.31	20.85
Medical expense—On major operation in the past 2-3 years	0.39	3.95	25.70	5.37	2,550.12	127.64
Communication (postage, phone, telegram, etc.)	14.90	41.92	81.58	129.97	323.29	46.57
Entertainment and leisure (cable charges, movies, etc.)	2.61	6.05	21.40	9.85	81.30	9.26
Expenses on social and cultural function	5.35	17.67	26.30	122.16	84.82	18.50
Expenses due to agricultural investment	39.23	180.17	446.19	1,076.75	5,947.91	436.77
Expenses due to litigation	3.95	23.63	66.25	91.17	4,857.09	253.30
Interest on loan	3.99	18.44	55.57	69.92	4,416.89	228.98
Premium paid on insurance	2.46	5.28	6.64	60.56	755.74	42.38
Any other (specify)	0.73	0.47	6.11	11.78	74.64	5.24
Total Monthly Expenditure (₹)	873.10	2,185.54	3,788.84	6,074.48	38,142.63	3,446.71

Strategically, the need for providing subsidy to this category of households appears to be absolutely essential. However, cash subsidy is unlikely to work because the households are likely to divert the money for other purposes. Subsidy should be in kind, having no resale value, and granted following a massive awareness campaign that use of toilet would reduce, in the long term, 10 per cent of the expenditure that households currently incur for medical purpose. Further, it is essential to explore if an insurance package could be designed linking construction and use of toilets with financial support against some basic health issues associated with open defecation. The premium could be paid from existing government schemes with a token contribution by the households.

Strategic Choices

MDG in sanitation for UP is achievable by taking a mission mode approach in respect to 22 identified districts and for converting unimproved toilet use into improved toilet use. While the former has potential to add 65 million new users, the latter can add as much as 33 million more.

It is essential to empower and incentivise *panchayats* for prioritising sanitation. Urban sanitation, particularly in respect to conversion of unimproved to improved, is also important to focus on. In rural areas, UNICEF may consider appointing District Coordinator (two for each of the 4 Group 1 districts and one for each of the 18 Group 2 districts). The presence of District Coordinator will ensure that sanitation is high in the political and administrative agenda at implementation level. The coordinator can also provide guidance on the implementation process.

As the analysis in previous section showed, toilet use does not increase because of expenditure. Hence, subsidy may not be the best mechanism to enhance sanitation coverage, although it does appear to be alluring and seems intuitively supportive. However, statistical analysis does not appear to support this expectation. While more research is needed for a firm conclusion, past experience along with current findings point towards the need to be cautious with expenditure-driven approach, including excessive dependence on subsidy. Where needed, subsidy should be direct, linked to output and in kind. Grant of subsidy should be preceded by awareness programmes.

Piped water supply is intimately linked to sanitation. Hence, sanitation programme should be planned in tandem with piped water supply programme of the state.

Link of sanitation with female literacy and BPL points towards the need to identify existing social networks of women, particularly in districts where BPL population is high, and utilise these networks for promoting sanitation. It is essential to identify Dalit concentrated districts among the 22 shortlisted districts and to prioritise them for inclusive growth. Further, fund requirement is both from the government and contribution by private individuals. In districts with high concentration of BPL, it may be essential to dovetail National Rural Employment Guarantee Act (NREGA) with sanitation planning because Dalits and low-income families may simply lack disposable fund for sanitation-related expenditure.

Net Social Benefit of Sanitation

The paper computes in rupees the net social benefits accruing to the society from sanitation. The benefit computation will enable advocacy with policy makers who often fail to appreciate importance of sanitation policy because the gains are indirect. Positive externalities have been computed using economic tools and techniques to explain the magnitude of benefits associated with sanitation.

In the current policy scenario, sanitation is pursued with limited political will, resulting in slow increase in the number of households using toilets each year. The increase in population is also adding to the burden of sanitation. Poor sanitation contaminates water which results in diarrhoeal illnesses. It is estimated that nearly 1,000 children below 5 years die from diarrhoea each day in India and 25 per cent of these deaths occur in UP.[2] Further, a large number of persons become sick and are unable to work resulting in loss of income. They also spend considerably in purchase of medicines.

The cost of total sanitation is the amount government and individuals will have to spend to have toilet in each household that the individual is committed to use. The total amount spent by government and individual, in average, for construction of toilet is ₹ 2,500 per unit. Each household will need a unit and currently there are nearly 25 million households without functional toilet.

---

2. *http://www.eddcontrol.org/region/asia.*

The benefits from sanitation are several and all are indirect. Proper sanitation results in lower pathogenic contamination of water, reducing the incidence of diarrhoea. Between 90,000 and 100,000 deaths each year (most of them children below 5 years old) are attributable to diarrhoea and these lives can be saved if total sanitation is achieved. The per capita income is ₹ 10,637 which is the average economic value of an individual. Computing for his entire lifetime, at a 10 per cent discount rate, the economic value of human life, in average, is ₹ 106,370. This is the benefit which an average individual can provide to the society of UP if he lives his full life. Obviously, this is a conservative figure. The reason for taking a conservative figure is to understate benefits and overstate costs, ensuring that positive net benefits, if any, are the most likely scenario.

Further, about 1 per cent of population fall sick from drinking contaminated water and such incidences can be reduced to 0.70 per cent by promoting toilet use. In average, a person is out of work for nearly 5 days each time he is sick and spends about ₹ 200 for treatment. These costs can be avoided if total sanitation is achieved.

If sanitation trend were to remain unchanged, deaths and illness described above would recur each year far into the future.

Taking a 10 per cent discount rate, the net present value (NPV) of net social benefit was computed using above facts and figures. The net social benefit was found to be ₹ 10,244 crore, which is nearly 4 per cent of the state domestic product. This is a conservative estimate as many other benefits of sanitation such as gains to tourism, saved deaths from other disease attributable to poor sanitation, etc., have not be taken into account for this computation.

Conclusion

The achievement of MDG in sanitation for UP requires a mission mode approach with focus on select districts where total sanitation should be pursued. Conversion of unimproved toilets to improved ones should also be a part of the overall strategy. Expenditure-driven (or supply-driven) approach is unlikely to work.

Female literacy and piped water supply positively impact toilet use. The net social benefit of total sanitation is nearly ₹ 10,244 crore which makes the goal worthy of early realisation.

13

RUSHABH HEMANI, RUNTI CHOUDHURY, CHANDAN MAHANTA, SOMNATH BASU and PARTHA PRATIM BARUA

Water Security Pilot Programme in the State of Assam

Gains and Gaps

Introduction

Threats to human health from water-related issues are linked to both scarcity and quality concerns (Moore, 2002; Zeman *et al.*, 2006). Worldwide several million people are plagued by lack of access to and availability of safe and secure water to meet their basic needs. The United Nations Children's Fund (UNICEF) and World Health Organization (UNICEF-WHO) (2012) Joint Monitoring Programme (JMP) report states that even though more than 2 billion people have gained access to improved drinking water sources and 1.8 billion people have gained access to improved sanitation since 1990, there are still over 780 million people without access to an improved drinking water source and 2.5 billion still lack access to improved sanitation.

With the established linkage between water quality, sanitation and health, UNICEF and WHO estimate that, at the current rates of progress, only 67 per cent—coverage of the Millennium Development Goal (MDG)—sanitation target will be reached by 2015 which, though better than previous projections, is still far from the 75 per cent coverage needed to reach the MDG target. This clearly indicates that unless the pace of change in the sanitation sector can be accelerated, the MDG target may not be reached until 2026 (UNICEF and WHO, 2012).

In terms of reaching the MDG on safe water, despite India being on track, it is significant that India and China together are home to some 216 million people, representing 28 per cent of the world's population, who continue to remain unserved and without access to improved water supply.

The provision of adequate and safe drinking water is a basic service that directly translates to health benefits and sustainable livelihood opportunities. WHO points out that $ 1 invested in water and sanitation translates into benefits worth $ 6 due to these multiplier benefits (Hutton and Haller, 2004).

National Rural Drinking Water Programme (NRDWP)—a flagship programme led by the Ministry of Drinking Water and Sanitation (MDWS), Government of India (GoI), has accorded greater emphasis to ensure drinking water security in rural areas. This is being done through measures to improve and augment existing drinking water sources and conjunctive use of water, based on village water budgeting and security plans prepared jointly by the community and the local government. The programme further aims at providing water security to communities by ensuring adequate and regular water supply, with quality and in conformity with prescribed standards at both supply and consumption points. The issues of potability, reliability, sustainability, convenience, equity and consumers' preference are the guiding principles that need to be taken into account while planning for any community-based water supply system.

The revised NRDWP guidelines in 2009 envisage communities to be at the core of monitoring and surveillance of drinking water sources. The guidelines further envisage providing an enabling environment for Panchayati Raj Institutions (PRIs) and local communities to manage their own drinking water sources and systems, including those in schools and at angawadi centres. Finally, to bring in transparency, accountability and informed decision-making, all information is required to be placed in the public domain.

Commensurate with change in the policy environment, assuring drinking water security and safety at the household level implies application of new and/or alternate technology, changes in approaches, practices and mindsets across the spectrum of stakeholders. In addition, the institutionalisation of village level management systems, through effective convergence between *panchayats* and communities that would facilitate decentralised planning, including village-level water budgeting, supported by adequate capacity building initiatives remain key to the water security programme.

The revised NRDWP guidelines envisage a renewed thrust on water quality monitoring and surveillance under which it is proposed that household level data on water security would be linked to a central database at the mission level. As a step in this direction, states are required to prepare district-wise drinking water security plans, which are derived from the village action plans (VAPs) prepared by the *panchayats*. This district level planning and coordination is extremely significant as it provides an institutionalised means of scaling up and strengthening village level planning, and linking it with district plans across areas of quality and quantity (MDWS, 2012).

Water Security Pilot Programme (WSPP) in Assam

Assam, the gateway to the northeastern region of India lies between graticules 89°42' to 95°16' E longitude and 24°08' to 28°09' N latitudes. With an approximate population of 31.17 million (Census of India, 2011),[1] the state covers a total geographic area of 78,438 sq km which is 2.4 per cent of total geographical area of India. Assam consists of 27 districts and 237 blocks, 2,580 *gram panchayats* (GP), 25,964 villages and 86,976 habitations (MDWS, 2012).[2] Almost 71.7 per cent of Assam lies within the 800 km long and 130 km wide Brahmaputra Valley, while the southern part of the state lies in the Barak valley, separated from the Brahmaputra valley by the Central Assam range. Nearly a fourth (720 km) of the Brahmaputra river's 2,890 km flows through Assam, descending 28 m from the Eastern Himalayas (Singh *et al.*, 2004), and is the lifeline of the people in the region.

1. *http://censusindia.gov.in/*, retrieved in June 2012.
2. *www.mdws.nic.in*, retrieved in June 2012.

Although there is abundant surface water in Assam, groundwater continues to play a significant role in meeting the water demands of most communities in the state, especially those in the rural areas. With 22 per cent of its groundwater resource being developed (MDWS, 2012), Assam continues to predominantly rely on groundwater as a source of drinking water, with tube wells and dug wells as the most common means to access water in rural areas. The Public Health Engineering Department (PHED) implements a number of Public Water Supply Schemes (PWSSs) that tap deeper aquifers with water treated through aeration and slow sand filtration.

Of Assam's total rural population of 26.4 million, 17.7 million (67%), living in 43,049 habitations, has so far been covered under the NRDWP. However, only 8.37 per cent of the population in the covered habitations contributes to the operation and maintenance (O&M) of the water supply schemes (MDWS, 2012), leading to the dysfunctional condition of some schemes that have been provided. Further, presence of iron and fluoride, bacteriological contamination and the relatively recent findings of elevated arsenic contamination in groundwater have raised a major concern regarding water security and safety issues in the state. As more and more areas are coming under threat of one or another kind of contamination, water security and safety is emerging as a major cause of concern in the region.

As mandated by the NRDWP, the Water Security Pilot Programme (WSPP) in Assam was launched in 2010-11 as a collaborative effort of PHED, Government of Assam, UNICEF and the Indian Institute of Technology (IIT) Guwahati, in partnership with field-based non-governmental organisations (NGOs), GP facilitators, *jal surakshaks*[3] and community volunteers. It was piloted in six districts *viz*., Dibrugarh, Golaghat, Jorhat, Kamrup, Sibsagar and Tinsukia, covering 12 blocks and 154 GPs (Figure 13.1).

3. *Jal surakshaks* are the volunteers for collecting water samples from the village and share the findings of the water sample analysis and messages on corrective action back to the concerned communities.

The prime objective of the programme was to demonstrate household and community level engagement in water security and safety initiatives, leading to the development of water security plans at the village level, with communities at the core of the planning process. To facilitate and institutionalise this process, PHED officials and NGOs intensively engaged with communities for the formulation and preparation of the VAPs, taking into account community know-how and practices relating to water sources. To strengthen the water quality aspect, the district level laboratories (DLL) were upgraded with high-end equipment/instruments.

Figure 13.1

Map showing the Six Pilot Districts along with the Selected Blocks under the Water Security Pilot Programme in Assam

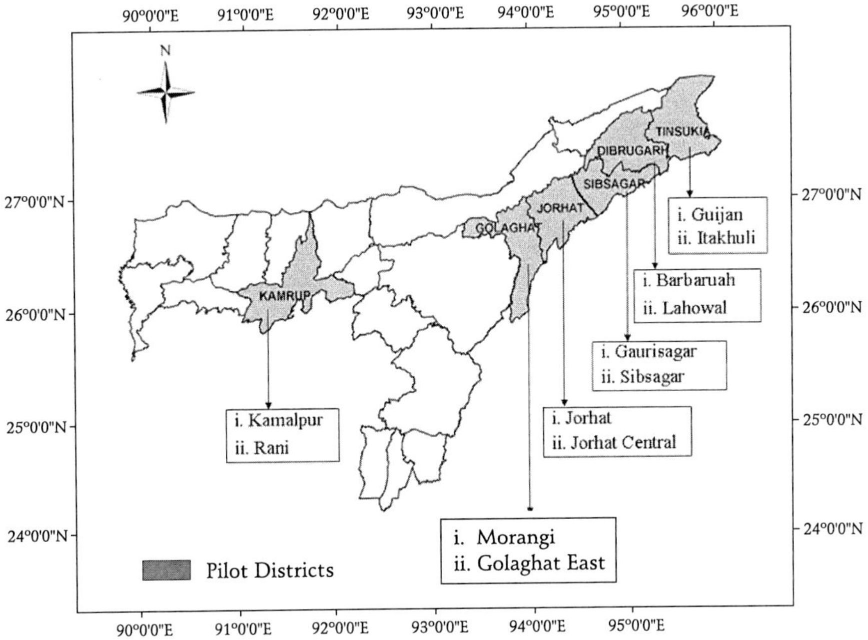

Note: Indicative map of Assam, not to scale.

Planning and Implementation of the Pilot

To facilitate planning and implementation, an effective institutional mechanism was developed which clearly defined the role and responsibilities of each key stakeholder *viz.*, PHED,

UNICEF, IIT Guwahati and the selected NGOs (Figure 13.2). At the district level, a PHED official was identified as the nodal officer responsible for monitoring and supervising the WSPP. A field-based NGO was also selected in each district as the interface between the PHED and the community.

At the NGO level, a district resource person, responsible for co-ordinating and monitoring the district-level activities, and one block coordinator for each of the selected blocks, responsible for block-level activities, were identified. The responsibilities of the block coordinator included compilation of the baseline data, collection of water samples for testing by the DLLs, and coordination with the field team, comprising GP facilators, *jal surakshaks* and community level volunteers. The field team was responsible for collection of baseline data pertaining to water samples, disinfection of water sources, sanitary surveillance, and community mobilisation and sensitisation.

As a preparatory activity prior to the implementation of the WSPP, two training manuals, one on Drinking Water Quality Analysis, Monitoring and Surveillance and the other on Capacity Building of Nodal Officers and NGOs on Water Security Programme, were developed and disseminated to the DLL personnel and other stakeholders including NGOs, respectively.

Concurrently, as mentioned earlier, the six existing DLLs in Dibrugarh, Tinsukia, Jorhat, Sibsagar, Golaghat and Kamrup districts were strengthened through both technical and managerial support that included capacity building of DLL staff, inventory management, quality control and verification, troubleshooting, and supervision. Necessary technical and managerial support for operationalising these laboratories were provided by the project team from IIT Guwahati, including required training, capacity building, inventory management, trouble shooting, supervision, quality control and verification, and confirmation of accuracy and precision of analysis.

At the village level, the *jal surakshaks* and the community level facilitators played a key role in disseminating water quality analysis results to the communities and suggesting disinfection or remedial measures to address contamination. To enable them to fulfill these responsibilities, they were provided intensive training on precautionary measures that were required to be followed during sample collection and preservation. These trainings were supplemented with information modules and manuals that highlighted the key elements of how sample collection and preservation is to be carried out.

Figure 13.2

Flow Chart of Institutional Mechanism Adopted for the WSPP in Assam

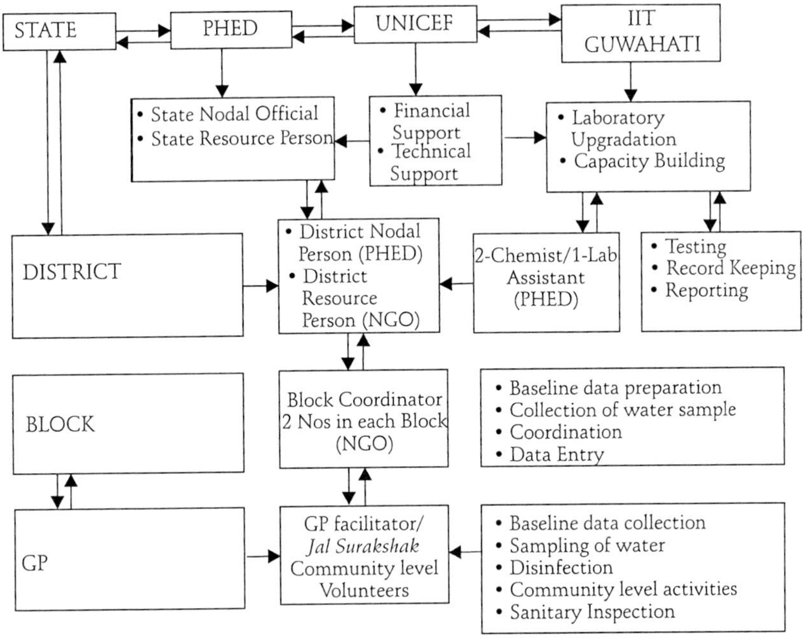

Preparation of Village Action Plans (VAPs)

VAP was prepared with active engagement of communities, which was at the core of the WSPP. The guiding principle of the WSPP was community involvement and improving quality of their lives by ensuring round the year adequate quality water for their drinking, cooking and hygiene purposes for all the communities.

The VAP preparation was a multi-pronged process that included various participatory techniques such as focus group discussions (FGDs), household and institution-level surveys using pre-tested questionnaires, transect walks, survey and mapping of status of household water sources including sanitary surveillance, water quality testing and the status of O&M of available water sources.

The VAP development process also took into account the current water availability and consumption pattern, existing schemes, identification of possible additional sources to augment availability of potable water, community perceptions regarding their water sources and quality, water demand, and measures to ensure sustainability.

By December 2011, a total of 980 against a target of 1,331 VAPs covering 154 GPs were completed. This was followed up by a state level review in March 2012 to discuss and decide way forward on the implementation of the VAPs.

Outcomes

To reduce the incidence of water-borne diseases, it is important that communities are aware of the importance of water quality and, in turn, adopt practices relating to the consumption of safe and potable water. Keeping this objective in mind, the WSPP advocated the integration of water quality monitoring and surveillance outcomes into the WSPP.

As part of the WSPP, six DLLs were upgraded into state-of-the-art facilities for water testing and analysis in the region, with necessary equipment such as phase contrast microscope, incubator, hot air oven, benchtop pH, electrical conductivity (EC) and total dissolved solids (TDS) meters along with supporting glassware, accessories and other consumables with support of UNICEF.

In addition, recognising quality analysis of water samples, database management, laboratory maintenance, record keeping and reporting is critical to the effectiveness of the WSPP: a total of 12 dedicated laboratory personnel (chemist/analyst and laboratory assistants) in the six pilot districts were also appointed by PHED as per NRDWP guidelines.

As a part of capacity building on water quality, a total of 30 laboratory personnel from the DLLs and sub-divisional level laboratories (SDLLs) were trained during the implementation of the WSPP.

To meet the water quality perspective, a total of more than 5,000 samples were tested for a range of physico-chemical and microbial analysis. For the Tinsukia District alone, a total of 1,145 samples were analysed for physico-chemical and bacteriological contamination of which 22.27 per cent of the samples were reported to have bacteriological contamination above the permissible limits.

Revenue Generation from Strengthened DLL in Dibrugarh

It may be pertinent to particularly share experience of the District Level Laboratory upgraded and facilitated at Dibrugarh, which was covered as a part of the WSPP. During the F.Y. 2011-12, dedicated efforts of the laboratory personnel and strengthened laboratory facilities were instrumental in generating revenue of around ₹ 1,00,000 through testing of private water samples from tea garden areas in the District. This has not only led to increase in the revenue generation, but has gone a long way in setting an example of bringing about a positive change in mindset of the PHED officials including the laboratory personnel. It is expected that this will be a performance to be ameliorated by the other DLLs

As an outcome of community involvement and ownership, 980 against a target of 1,331 VAPs covering 154 GPs were completed by December 31, 2011, with communities at the core of the planning process. This entire process of VAP development has had a two-pronged positive effect; on the one hand, it has developed a cadre of trained and sensitised *jal surakshaks* and community volunteers and, on the other, it ensured enhanced community knowledge, demand and ownership of water security and safety issues.

Based on the WSPP pilot in the six districts, during 2012-13, the PHED has taken a decision to expand the WSPP to nine additional districts *viz.*, Barpeta, Bongaigaon, Cachar, Dhubri, Goalpara, Hailakandi, Kokrajhar and Nalbari.

Field Challenges

Despite initial teething problems and inherent challenges, WSPP has been an important milestone to demonstrate community involvement in the planning process for water security and safety in Assam. The field challenges included lack of awareness among communities on the importance of water safety and their lack of trust on government functionaries. Issues such as delays in release of funds from state to district and block levels, that led to non-availability of survey formats, chemicals, etc., coupled by high turnover and limited capacities of *jal surakshaks* had a negative impact on the overall implementation.

At the village level, defunct water supply schemes on account of poor O&M as well as lack of ownership led to non-cooperation by the community members during the WSPP planning process. For instance, it was observed that lack of availability of spare parts for Tara and Singur hand pumps often led to poor response by the community. There was also a clearly felt need for the comprehensive capacity building of *panchayat* representatives to enhance their ownership of water security projects.

From the geographical perspective, remote and inaccessible habitations, particularly during the monsoon season, remained a challenge that was faced during the field survey.

Further, critical issues such as remuneration of laboratory staff, insecure contractual modalities, non-repair of critical instruments, non-utilisation of laboratory equipment, timely replenishment of reagents, intermittent power supply, etc., hindered the effective functioning of the laboratories.

Lessons Learnt

Implementation of the WSPP resulted in several pertinent and key learnings, which, if addressed, would go a long way in strengthening the implementation of similar interventions in future. Some of the key lessons learnt are summarised below.

Systemic Issues

For the effective implementation of WSPP, it is imperative to address systemic issues that have hindered field and laboratory functioning, viz., timely release of funds, timely and quality logistics (survey formats, chemicals/reagents, equipment), poor

O&M of schemes, staffing issues (remuneration, contractual modalities), intermittent power supply, etc.

To address these systemic issues, it is critical to have in place an established protocol on 'Standard Operating Procedures (SOPs) for community based water and security programme' which would facilitate the effective implementation of the water security programme in the state.

Community Interface

The key objective of the WSPP was to encourage and involve communities in the preparation of the VAPs through participatory dialogue and processes. Recognising that most often it is the women of the household who are primarily responsible for fetching and ensuring water security, special efforts were made to sensitise them and bring them on board during the VAP preparation process. During the FGDs, more than 50 per cent of the representatives were observed to be women participants. This strategy needs to be reinforced while implementing similar interventions in future.

Experience has shown that decentralised governance is a core element for the successful execution of the WSPP, with focus on O&M. With training and awareness-generation, it is feasible that communities are mobilised to collect water revenue and maintain schemes. The option of local kiosks such as rural marts or other dependable sources for procurement of spare parts and other supplies need to be explored so that communities are equipped to undertake O&M activities.

Community awareness on the critical linkage between water quality, sanitation and health outcomes was often found to be lacking, indicating the need for a more comprehensive awareness generation on the importance of water safety at source and up to consumption at the household level.

Way Forward

With the completion of the WSPP in the six districts of Assam and with a total of 980 VAPs in place, there is a clear need to evolve strategies for the successful implementation and sustainability of the VAPs. Keeping in view the formidable challenge of addressing water security and safety for all water disadvantaged communities, this pilot programme was able to achieve a relatively small but vitally important first step towards involving communities in the planning of village level water security plans. The findings, challenges and lessons learnt from this pilot initiative are expected to facilitate better understanding of field realities and to contribute to the refinement of the planning and implementation process for water security programmes elsewhere in the state as well as at the national level. There is also a felt need for more in-depth analysis and introspection of behavioural and management challenges, especially at the systemic level, which could lead to peer learning among different districts as well as states.

Despite apparently adequate water resources, Assam continues to face quality-related challenges for the provisioning of safe water especially to vulnerable rural communities. Considering inherent weaknesses such as below-average income, social marginalisation including the inherent low priority accorded to the linkages between water-sanitation-health, and overall poor water governance, it is critical to develop strategies and action plans that rely on good decentralised governance systems.

A comprehensive review of the 980 completed VAPs and formulating strategies for their effective implementation is the need of the hour. Such a step would help strengthen the trust and confidence between the key stakeholders, *viz.*, the government and the community, contribute to sustainability of interventions, and pave the way for accelerated programme implementation in other districts.

At present, any revenue generated by either the state, district or sub-district laboratories gets deposited in the State exchequer without benefit to the laboratory concerned. For laboratories to function to their optimum potential and retain the motivation of staff, a mechanism needs to be institutionalised to ensure that at least part of the revenues generated is ploughed back to the respective laboratory towards its upkeep/maintenance. This calls for a policy level intervention including directives for creating a separate bank account to deposit and re-deploy the revenues generated by these laboratories.

Based on the findings, challenges and lessons learnt, the following recommendations have been discussed and endorsed at a state level review of the WSPP, chaired by Secretary, PHED, with participation of all key stakeholders.

These recommendations have been broadly classified into three areas *viz.*, implementation of water security plans/VAPs, strengthening of water quality laboratories and community involvement.

Implementation of Water Security Plans/VAPs

- With substantial allocations (₹ 132 million) released by the MDWS to Assam for water quality monitoring and surveillance, there is an urgent need to formulate policies on how best to use this allocation to support and implement the VAPs.
- As an outcome of the WSPP, 980 VAPs have been prepared involving communities as key stakeholders in the process. The challenge now is to 'convert' these VAPs into detailed project reports (DPRs) against which water supply schemes can be sanctioned in line with the NRDWP guidelines. This is critical to ensure that efforts made to involve communities and promote their ownership for water security and safety get the required buy-in at all levels of government.
- Field implementation of VAPs needs to be carried out by establishing effective convergence with relevant sectoral departments including Panchayats & Rural Development (P&RD). Capacity building of PRI members is key to this process to ensure that district administration and PHED are empowered for the effective implementation of the VAPs.
- As the Action Plan on Water Sector for 2012-13 has been prepared for Assam without budgeting for the implementation of the VAPs, a revised action plan covering at least 10 VAPs per pilot district (i.e. in total 60 VAPs) in

line with the NRDWP guidelines needs to be formulated on a priority basis.

- The district water and sanitation committees (DWSCs) need to be empowered to take decisions and implement the VAPs in each of the six pilot districts and, subsequently, in the nine newly identified districts.
- All preparatory activities including laboratory strengthening, logistics, data entry and analysis systems, streamlining and timely funds flow mechanism, transparent procedure for NGO selection, implementation plans, and monitoring mechanisms need to be put in place for Phase II of the Water Security Pilot in the nine newly identified districts. A peer-to-peer learning mechanism also needs to be established by linking the nine new districts with the six pilot districts.

Strengthening of Water Quality Laboratories

- Stringent and strategic policies on water quality monitoring, that includes monitoring and surveillance of contaminated sources, and ensuring alternate sources for affected communities need to be developed.
- Based on a review of the performance of the upgraded DLs, the existing laboratory protocols would need to be revised so that a systematic approach for water quality monitoring and surveillance is put in place.
- With water quality as a priority area for the GoI, the smooth and effective functioning of DLL and SDLL needs to be ensured. As a step in this direction, some concrete action to retain the existing trained water quality laboratory technicians is required. This could include rational revision in their remuneration, streamlining contractual modalities, etc.
- For effective functioning of the DLL, it is important to meet the day-to-day expenditure through revenues generated by the respective laboratories. This will serve as an incentive and improve laboratory functioning, including operation

and maintenance. A policy level decision needs to be taken in this regard.

Community Involvement and Ownership

- Community awareness on the critical linkage between water quality and health is at the core of any community-managed water quality programme. Hence, capacity building of community-level stakeholders needs to be taken up.
- The traditional practice of infrastructure development for water security needs to be reviewed and greater emphasis on a user-driven water supply system needs to be promoted.
- Policies to involve communities in the maintenance of their alternate water sources as well as the piped water supply systems need to be formulated and implemented. This calls for setting up of users' groups and self-help groups (SHGs) to manage water sources.
- Fostering higher participation of women and disadvantaged groups in water supply interventions and water user groups, particularly women heads of households, can be rewarding as women are primarily responsible for fetching and ensuring water security at the household level.
- To promote the sustainability of rural water safety and security interventions, it is important to facilitate greater involvement and build capacities of organisations that support communities (NGOs) and the local private sector.

Conclusion

The WSPP piloted in Assam was an initiative to involve communities as key stakeholders in the water security planning process in alignment with NRDWP guidelines. This pilot programme showed that community involvement right from planning processes contributes to enhanced ownership not just in terms of physical resources but also in terms of the programme. With the outcome of this pilot programme in hand and community perceptions known, the further challenge of implementation of the 980 VAPs covering 154 GPs has been recognised. Apart for the VAPs, laboratory personnel in the five DLL equipped and upgraded, actively participated in the water quality monitoring and surveillance, another important component of the pilot programme. Their involvement in the whole programme helped demystify some of the pertinent issues hindering smooth functioning of the laboratories.

With both encouraging and some downbeat experience at various levels of implementation, the broader lesson learnt was that a bottom-up decentralised approach involving communities and village level volunteers backed by stakeholders at all the levels can indeed ensure successful implementation of the water security programme irrespective of challenges involved.

Reference

Hutton, G. and L. Haller (2004). "Evaluation of the Costs and Benefits of Water and Sanitation Improvements at the Global Level". Retrieved from the url: *http://www.who.int/water_sanitation_health/wsh0404.pdf*

MDWS (2012). Retrieved from the url: *www.mdws.nic.in*

Moore, G.S. (2002). *Living with the Earth: Concepts in Environmental Health Science*. 2nd ed. New York: Lewis Publishers.

Singh, A.K. (2004). "Arsenic Contamination in the Groundwater of North Eastern India", in: *Proceedings on National Seminar on Hydrology*, Roorkee.

UNICEF and WHO (2012). "Progress on Drinking Water and Sanitation". *JMP Report*. 2012 Update, Retrieved from the URL: *www.unicef.org/media/files/JMPreport 2012.pdf*

Zeman, C., M. Rich and J. Rose (2006). "World Water Resources: Trends, Challenges, and Solutions". *Reviews in Environmental Science and Bio/Technology* 5:333–46.

Disclaimer: The views expressed herein are those of the authors and do not necessarily reflect the views of the United Nations.

14

SONALI SRIVASTAVA

The Gram Panchayat Organisation Development Project (GPOD)

Introduction

We are well past the discussion of whether the *gram panchayat* (GP) should be a strong institution or not. A strong GP is the solution for solving the long-lasting service delivery issues to the last mile, the rural population. A strong and well functioning GP can make a significant improvement in local self-governance and delivery of various rural development programmes, targeted at 60 per cent of India's population. And anyway, it is mandated by the 73rd Amendment to the Constitution of India.

There are many efforts to strengthen the GP and its importance, including assigning funds, allocating funds and exploring methods to assign manpower to the GPs. In addition, each state has a State Institute of Rural Development (SIRD) to provide training to the GP members and staff. At present, most institution and capacity building efforts are focused on class room training.

However, despite two decades of such efforts, the decentralisation agenda has not really taken off, primarily as only few *panchayats* have risen to their mandate and role. At this juncture, we need a paradigm shift in the manner we look at GPs. Examining the roles passed on to GPs in various programmes and schemes by government departments, one can observe that GPs are seen more as an extended service delivery arm of the government in a top-down development framework and not as units of self-governance. To be able to leverage resources from other government institutions, triggering higher accountability on one hand, and more importantly, better service delivery and governance on the other, the GP has to be a strong organisation.

The GP Organisation Development (GPOD) project was conceived to look at capacity building from an organisational perspective, trying to address one fundamental question: how to strengthen the GP organisation and bring about greater convergence at the GP and its associated departments, with the aim of improving governance and service delivery to rural citizens? The project was executed at two levels. At the first level, it was to understand, observe and critique the design and functioning of institutions involved in governance and service delivery to rural citizens. At the second level, with a view to test our various hypotheses as well as to evolve, pragmatic solutions which are implementable in the real context, we worked with two GPs in Karnataka, facilitating real time change in an action research mode.

The framework of the GPOD project is based on organisation development (OD) principles. The rationale for the using OD framework is its focus on organisation as a whole, and not on standalone components. The premise is that successful change occurs when all components of an organisation are aligned, i.e., the vision, skills, incentives, resources, action plan and results. Diagnosing that there was lack of structural and functional clarity, we worked extensively with two *panchayats* to develop key foundational principles. (i) We believe that the the GP is pivotal in improvement of service delivery and governance at the last mile. To be able to perform mandated functions, the present capacity building efforts are inadequate. (ii) The focus should be on building a strong GP organisation, which has higher absorptive capacity to be able to leverage different efforts and resources to improve the GP. (iii) Elected representatives can potentially contribute significantly in enabling the GPs perform their roles. (iv) Lastly, a number of policy changes have to be taken to create an enabling environment for the GPs to function as an equal intergovernmental body that it is.

This paper dwells on the project methodology, our findings against the hypotheses that we started with and our recommendations.

Background

A seed was sown with A Survey of Household Water and Sanitation (ASHWAS), which was conducted by Arghyam[1] in 2008, in 172 GPs across 28 rural districts of Karnataka. Some of the key findings are detailed in Box 14.1. Post survey, dissemination meetings were held with all participating GPs, which were half day meetings with the Gram Panchayat members, Executive Officer of the Taluka, invited community organisations and citizens. We presented the findings of the survey and sought a discussion among participants on how they would like to address these issues. We hoped that these presentations and GP specific reports, which were handed over to the Sarpanch, would act as strong catalysts for planning and implementation towards addressing issues emerging from ASHWAS.

BOX 14.1

ASHWAS: Key Findings

Water

- For most people, public tap is the primary water source followed by piped water & mini water supply.
- Nearly 22 per cent of the citizens do not get water throughout the year from their primary source.
- 45 per cent households face water shortage for more than a month.
- Per cent levels beyond permissible limits for water quality
  - Fluoride 60 per cent (at WHO standards)
  - Bacterial 38 per cent
  - Nitrate 20 per cent

contd...

1. Arghyam is an Indian public charitable foundation set-up with an endowment from Ms Rohini Nilekani, and working in the water and sanitation sector since 2005.

...contd...

Sanitation

- 72 per cent of the citizens practice open defecation.
- 75 per cent citizens either do not have access to drains or their drains are never cleaned.
- 63 per cent people do not undertake any form of treatment before drinking water.

O&M and Governance

- Only 21 per cent GPs in Karnataka used the water quality testing kit provided free by the government. It is another matter that most of these kits have either not reached the GPs or are lying unopened.
- 73 per cent disruption is due to lack of O&M, with only 27 per cent being due to natural causes.
- Only 42 per cent of stipulated Watsan committees (Village Water and Sanitation Committees—VWSC) were active in the GPs surveyed.
- ₹ 14/- was collected per capita per annum as water tax, as against ₹ 46/- per capita expenditure on Watsan. On an average, 10-20 per cent households pay water tariff.

However, most GPs did not have the capacity to internalise these issues and take action. The recurrent impediments articulated by members were lack of funds, no government support, corruption, lack of knowledge, etc. The mood in these meetings ranged from indifference to despondency and, in some places, anger with the present state of affairs. In a very few GPs, there was optimism and action orientation. In other words, they did not look like organisations which could take over effective execution of multiple government programmes. Some of the organisation and finance-related issues are detailed as follows:

Organisational issues	*Financial*
• Responsibilities laid down for GPs far outweigh their knowledge and organisational capacity to fulfill them.	• Lack of awareness regarding budgets
• While many programmes rely on GPs for implementation, they have limited powers/ authority, leading to diffused identity.	• Lack of control over funds
• Poor incentives to perform: GP members are expected to dedicate their time without being compensated.	• Insufficient revenue generation (10-20% population pays taxes)

Given the above context, we delved further to understand issues related to capacities of panchayati raj institutions (PRIs).

Efforts for increasing capacities of GPs vary and can be detailed as:

- Programmes to strengthen leadership, where the *sarpanch* has taken up the mantle to improve the GP functioning as a whole.
- Government or at its behest, NGOs, co-opting and training GPs or their sub-committees for the implementation of specific programs, such as Integrated Watershed Management Programme, Sachetana (rainwater harvesting), etc.
- Targeted programmes such as Backward Regions Grant Fund (BRGF) and Gram Swaraj (World Bank funded Karnataka programme), which aim at supplementing developmental flows to PRIs, with conditionalities on processes to be followed such as participatory planning, adherence to accounting norms, etc.
- Training programmes by various SIRDs (State Institutes of Rural Development) targeted at the executive and elected members of the PRIs.

Prior to initiating the project, several efforts were taken to understand the present situation of the GPs and views of experts in the field:

- Understanding of GPs' performance and the challenges faced by them through ASHWAS and an Asian Development Bank (ADB) study.[2]
- Understanding of government programmes from various documents and visits to GPs covered under different programmes.
- Reference to existing legal framework—Karnataka Panchayat Raj Act, 1993, accounting and budgeting guidelines, etc.
- A further understanding was also gained from visits and discussions with authoritative individuals and organisations:
    - o Dr Mihir Shah, Member, Planning Commission.
    - o Mr Ravikumar, Secretary, Rural Development and Panchayati Raj Department, Government of Karnataka.
    - o Mr T.R. Raghunandan, Ex Indian Administrative Service (IAS) Officer, now working actively in GP capacity initiatives
    - o Shri V. Ramachandran, Retd. IAS officer, member of several bodies such as Administrative Reforms Commission, Institute of Rural Management Anand (IRMA), National Dairy Development Board (NDDB), etc.
    - o Mr R. Elango, Managing Trustee, Trust for Village Self Governance, Kuthambakkam, Tamil Nadu. We met Mr Elango at the model Kuthambakkam GP.
    - o ANSSIRD (Abdul Nazir Sab State Institute for Rural Development), Mysore.

2. TA-6325 (REG): Promoting Effective Water Management Policies and Practices—Phase 5, conducted on behalf of ADB and UNESCAP.

- o Gram Panchayat Nademavamapura—Kunnigal *taluka*, Tumkur, covered under Gram Swaraj Scheme.
- o Gram Panchayat Vasana—Davangere district, and the District Training Centre, run by SIRD, Davangere: covered under Backward Regions Grant Fund (BRGF).
- o Gram Panchayat Hebballi, Chitradurga District, where remarkable work has been done by the Village Water Sanitation Committee (VWSC) towards water management, initiated under project funded by Danish International Development Agency (DANIDA).
- o Analysis of 10 strong and weak GPs conducted by communications, development and learning.
- o Myrada.
- o Department of Watershed, Government of Karnataka (GoK).
- o Sachetna Fluoride mitigation—Programme officer.
- o Consultation at Planning Commission on November 24, 2010, with NGOs, elected representatives, members from Planning Commission and Ministry of Panchayati Raj, Government of India (GoI).
- o Series of discussions within Arghyam.

Assessing Problems

The problem starts with poor delivery of basic services to rural citizens. The state has designed multiple programmes and schemes to deliver services to rural citizens. And yet, as the ASHWAS results demonstrated, the state of services is sub-optimal. Each programme lays down elaborate institutional structures and there are dedicated funds.

The PRIs, specifically GPs being the last mile organisations, are increasingly being seen as substitutes for the grass-roots delivery organisations. This reconstitution of functions has raised many pertinent questions: are GPs an implementation arm of the government or are they an autonomous entity which should decide and set their own priorities, using government machinery and funds selectively, as per their own needs.

The idea of 'decentralisation' as outlined by the 73rd Amendment is meant to serve a larger purpose of strengthening political and social identity at the GP level, while improving service delivery could be seen as one of its key goals, but not the overall encompassing agenda.

Therefore, to define the problem, we were trying to address: How do we strengthen the GP as an institution of self-governance, as constituted under Article 243, for the rural areas?

Project Objectives and Scope

The project was executed at two levels. At one level, it is to understand, observe and critique the design and functioning of GPs towards governance and service delivery to rural citizens. At the second level, with a view to test our various hypotheses as well as to evolve solutions which work in the real context, we are working with two GPs in Karnataka, facilitating real time change in an action research mode.

Accordingly, the project objectives were:

- to develop a step by step framework for developing a strong GP organisation;
- to initiate a process of real time change in two GPs in Karnataka while implementing the above framework; and
- to leverage local strengths and capacities to establish role clarity, better convergence and interaction with the government agencies.

As the GP does not function in isolation, roles of interfacing institutions were also being analysed. The process of the project execution has been to engage with GP members as partners and decision makers in problem identification and solution building, rather than them being recipients of change, this being the core principle of decentralisation.

Project Phases

The project was being implemented over a span of 2 years, spread over three phases:

I. Preparation phase

II. Diagnosis and design

III. Handholding implementation in two GPs

Preparation Phase

Key aspects of this phase were:

- Finalisation of scope and budget.
- Identification/selection of team and partners.
- Memorandum of Understandings (MoUs) with different consultants and partners of the project.
- Briefing external agencies which will play a critical role in the project such as Ministry for Rural Development and Panchayati Raj (RDPR), Karnataka and ANSSIRD.
- Setting up the project management structure.
- Selection of GPs.
- MoUs with the GPs.

Diagnosis and Design Phase

This was to understand the factors which impact the organisational effectiveness of a GP. As the GP does not function in isolation, the study included analysis of formal government organisations as well as informal organisations impacting the GP socially and politically. The output of the diagnosis was used to design the framework which could enhance organisational effectiveness and build organisational capacity of the GP. The key outputs of this phase were:

Diagnosis

- Basic information of the GPs, demography, CBOs and 61A committees, revenue generation, etc.

- Documentation of information collected regarding the present functions being performed by the GP: activities, people/agencies involved and their roles, and issues facing the GP. Process mapping tool was used to have discussions with the GP members, along with staff and key resource persons from the villages. Twenty such functions were mapped as processes to give significant insights on the present state of affairs in the GP.

Design

- Vision and purpose of the GP.
- Organisational structure of the GP.
- Capacity building methodology for members and staff.

Handholding Implementation in Two Gram Panchayats

This phase was designed to confirm the diagnosis and design variables of the project, while facilitating real time change in the selected GPs. Based on the action research during this phase, the overall recommendations for strengthening GP capacity would be finalised.

Accordingly, the objectives of this phase were:

- to facilitate the GP in setting up its organisational capacities to work towards its vision and purpose, as per the recommendations in the diagnostic and design phase;
- to work with the GP to move towards its vision and purpose, utilising its own capacity as well government machinery, schemes and resources as per need; and
- to simultaneously critique the processes and institutions laid down by the state in its interaction with the GPs.

Deliverables from the handholding phase:

- Implementation of the organisational design and linked human resource systems for the GP.

- Facilitating GP to develop the required annual plan(s), as well as develop detailed project and resource plans for implementation of the annual plan(s).
- Facilitate monitoring and implementation of the annual plans.
- Skill and knowledge building, and helping GP members identify solutions for different issues facing the GP while implementing the annual plans.

Hypothesis and Findings

We worked on three hypotheses:

1. We had hypothesised that extensive framework is being created by the state for improving governance and service delivery, decentralisation being one of the key steps. Implicit in this hypothesis is that there is a strong legal, financial and institutional framework to strengthen decentralisation.

   Based on our understanding over the past year and a half, we conjecture that while there is a legal framework, it is not being used in a manner which is conducive to decentralisation. In fact, based on this framework, workload on the GPs is continuously increasing due to pressure by the two key stakeholders of the GP—the citizens and the state. On the other hand, while the GPs have been given statutory powers to enable them to meet the above mandate, they are insufficient for performing as institutions of self-governance, which can negotiate and enter into contracts with different government and non-government institutions to improve service delivery and governance.

- As per the Karnataka Panchayati Raj (KPR) Act, 1999, the three tier Panchayati Raj system is "for greater participation of the people and more effective implementation of rural development programmes and to function as units of local-self-Government."

- To meet the first mandate, the KPR Act builds a strong relationship between the citizens and the elected members in the GP, who are representatives of different wards. Chapter III, KPR Act, details the powers and functions of the ward and *gram sabhas*, the latter being vested with the power to "consider and approve the annual plan prepared by the Gram Panchayat." The citizens have, therefore, been given the power to set priorities and approve the development agenda of their villages.

  The GP, on the other hand, is the organisation which has the closest proximity to people, each ward member representing the four hundred population of his/her ward. This proximity has ensured prompt communication of grievances and requirements of citizens, putting tremendous pressure on individual members.

- To meet the second and third mandates of effective implementation of rural development programmes and function as units of 'local self-government, we will analyse the often discussed framework of 'functions, funds and funcationaries.'

  The GP has been given many roles and functions as per law: 29 functions as per Schedule I, KPR Act, and, specifically, 17 functions which the GP has to mandatorily provide for, as long as funds at its disposal will allow (Section 58 1(a), KPR Act).

  Our conjecture is that this mandate has legitimised to some extent, the state's abdication of last mile responsibility of service delivery to the GP. Mostly, central government programmes now have an elaborate role for the GP, each of which may, in fact, need dedicated specialised organisation for an efficient delivery. Let us take the Mahatma Gandhi National Rural Employment Guarantee Act (MGNREGA) to illustrate this.

MGNREGA Guidelines

The village, intermediate and district *panchayats* shall be the principal authorities for planning and implementation of the scheme made under the MGNREGA.

- The *panchayats* at all levels can be the implementing agencies under the Act.
- At least 50 per cent of the works in terms of cost will be allotted to GPs for implementation.
- The GP shall be responsible for identification of the projects to be taken up in its area under the scheme as per the recommendations of the *gram*/ward *sabha* and the same shall be forwarded to programme officer for scrutiny and preliminary approval.
- The GP shall prepare a development plan and maintain shelf of possible works to be taken up under the scheme as and when demand for work arises.
- The intermediate *panchayat* shall approve the block level plan and forward the same to the district *panchayat* for approval.
- The district *panchayat* shall finalise and approve block-wise shelf of projects to be taken up for implementation under the scheme.
- The plan approved by district *panchayat* will assign implementation responsibilities to various agencies like *panchayats*, line departments, NGOs, etc.
- On the matter of funds, the GP is empowered to levy taxes and tariffs on citizens as well as factories/business enterprises, to raise revenue for improved service delivery (Section 199, KPR Act), including recovery of payment in case of non-payment (Section 200). In addition, there are funds transferred by the central and state governments to PRIs for implementation of various programmes and schemes.

However, some analysis on the key source of funds reveals a very different reality. As per the well-being rankings conducted as a part of partcipatory rural appraisals (PRAs) in Dibburhalli GP, Karnataka, 72 per cent of the population, which is the key resource base for raising revenues, comes in the categories of poor and marginalised. Therefore, tax collection from citizens suffers due to the low levels of income in rural areas.

In most GPs, there are not many factories and business enterprises. Finally, while funds transferred from the state are for 'implementation,' there are no untied funds under programmes for institutional expenses of the GP. Even where there are dedicated administration funds, such as under MGNREGA, they are retained at the state level.

- Regarding staffing, there are several provisions/existing resources which a GP can leverage:
  - Provision to employ additional staff at the GP, and to engage technical staff not belonging to the state (Sections 113 and 113A).
  - Three standing committees (Section 61) and provision to form sub-committees under Section 61A to 'perform' functions of the GP.
  - Elaborate set of line departments for delivery across functions of the GP.

But here, too, there are gaps. In addition to its answerability to the ward/*gram sabha*, GP has to manage the *thana* or *zila parishad* (TP/ZP). It has to take approval from the TP for staffing pattern, scales of pay and mode of recruitment of even those paid by GP. Also, the GP needs to meet conditions of reservation while hiring staff (Sections 112 and 113, KPR Act[3]). The TP and ZP have overriding rights to requisition and recover moneys, accounts, records, or

3. Accordingly to circular from RDPR, the permission is actually to be granted by the chief executive officer (CEO). There is also a bill collectors' union which prevents GP from taking action against an errant bill collector.

other property pertaining to a GP or *taluka panchayat* (Sections 157 and 198).

The two full-time staff of the GP, the *panchayat* development officer (PDO) and Secretary, while answerable to the GP as per the Act (GP *adhyaksha* is the 'executive head' of the GP as per Section 62), are paid by the government, and therefore have allegiance to the state.

The government has the power to specify the role of *panchayats* in respect of the programmes, schemes and activities related to the functions specified in Schedule I, II and III for their effective implementation (Section 240).

Given the controls and requirements as per the Act, the treatment of the GP is more like an extended delivery arm of the government. It is another matter that the Act does not formalise this by proposing remuneration to members for fulfillment of expectations.

2. Our second hypothesis was that implementation of the frameworks is constrained by inadequate organisational capacity of the GP. As stated above, the treatment of the GP by the state is at best like a government arm for implementation of rural development schemes. Most of the capacity building efforts are aimed towards providing them knowledge, skills and even tools for better implementation of government programmes, rather than building an institution which can decide its priorities and strengthen its resources and capacity for self-governance.

   A strong institution needs to have well articulated and commonly understood vision and mission, organisation structure and skills, laid down processes and systems, resources and finally incentives to perform. During the diagnosis stage, we found these key elements either absent or at best inadequate. Moreover, all efforts related to capacity enhancement are towards standardising processes in all GPs and ensuring uniformity in information collection rather than building a vibrant organisation.

GP's long-term goals, articulated in our project through vision and mission, are a case in point. A GP is to have multiple 'Perspective plans,' one each for RDPR, Finance Commission and MGNREGA respectively, aimed at money which can be received under these heads. Each of these is a five year plan which should be broken into annual plans. Should an institution have three sets of visions? The need is to develop a long-term vision for the *panchayat*, and evolve annual plans based on that. Budgets could also be integrated, including government, non-government and own funds.

The present GP structure is primarily a legislative structure, aimed at citizen participation. Access and understanding of funds and functionaries for service delivery from the state among GP members is low.

Figure 14.1

Gram Panchayat Structure

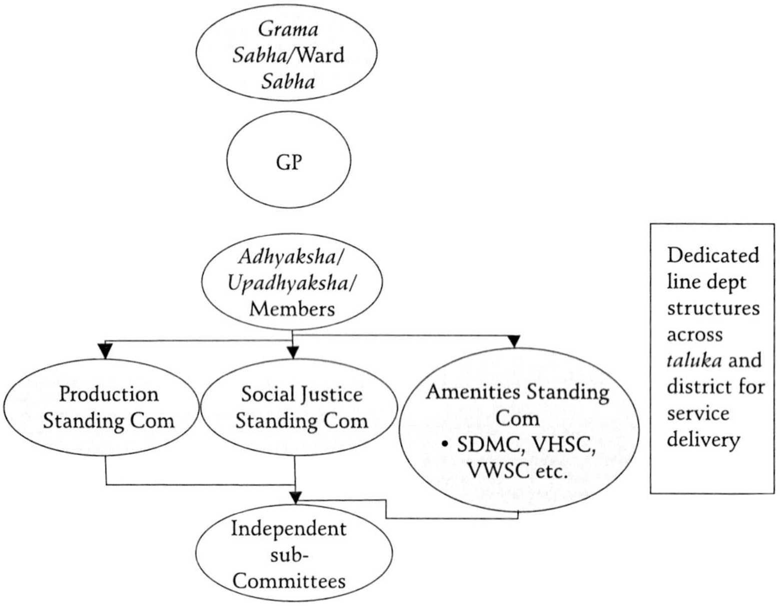

With members assuming responsibility, the service delivery structure looks something like as in Figure 14.2.

Figure 14.2

Service Delivery Structure

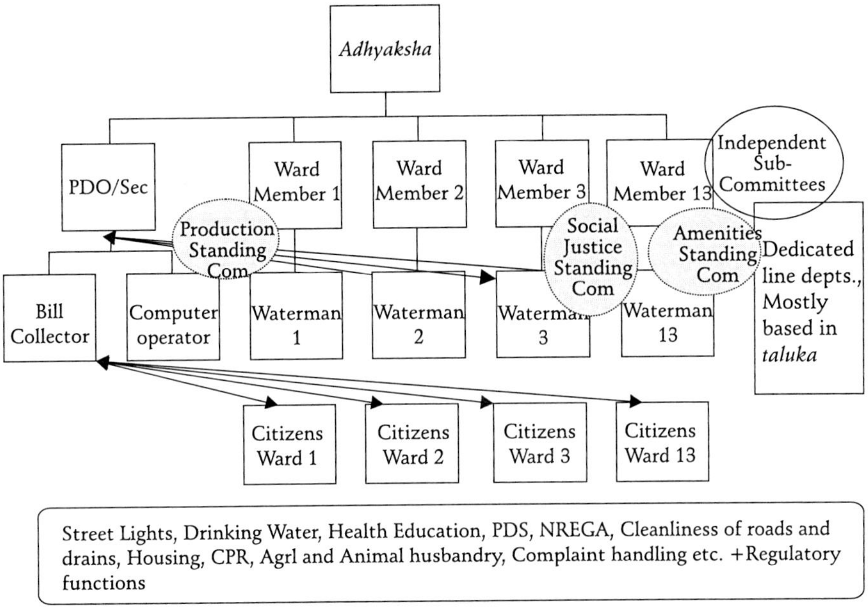

There are several issues with the present organisational set-up:

- Responsibilities and capacities of GP members are not aligned
  - o Even though line departments exist, the citizens hold GP members accountable for most functions.
  - o Responsibilities of government line departments, their roles, financial and manpower resources are not clear to the GPs, which leads to sub-optimal leveraging.
  - o With very few members possessing required knowledge, capability and motivation, most wards suffer as members underperform.
  - o Information and knowledge on regulatory powers are very weak.
- There is a lack of well laid down systems and processes to aid decision making. Most problem-solving in a GP is done through fire fighting.

- Planning and budgeting are done in hindsight. This results in GP's power structure taking precedence over equal representation.
- There are only two full-time staff (PDO and Secretary), who are government employees, and may have different allegiance and priorities. The government has earmarked Secretary for office work and PDO for service delivery, per GP. The PDO, while potentially a useful position, has been facing many challenges since its introduction a year back.
- GP is free to hire its own staff, but has to pay their salaries from its own funds. As a result, a skeletal staff exists, who may not have adequate skills and knowledge.

3. Our project was based on our third hypothesis that additional organisational and systematic insight of functioning of institutions will add to the existing efforts for improving service delivery and governance, aiding the decentralisation process. Accordingly, we worked with the two GPs on developing their vision and mission, understanding the current processes while detailing the RACI (responsibility, accountability, collaboration and information providing roles of different stakeholders involved in each process) and eventually developing an organisation structure which will help them in governance and service delivery. Throughout the exercise, the core methodology of decision making, i.e., through citizen participation, was followed which itself has significantly contributed to institutional capacity building. The step-by-step organisation development approach is detailed in the next section.

Components of GP Organisation Development

The key components of organisation development are detailed below, which were implemented in the two project GPs:

Vision and mission of the GP: Developing the vision and mission of the GP was an intense process, conducted through a two-day workshop, attended by all GP members, selected staff and additional resource persons such as school teachers, advisors, etc. The key input to the vision and mission were pictures drawn by participants of their dream village. What kind of values would they like to build in their GP? What kind of amenities would they like to see in their village, if resources were not a constraint? This exercise was empowering because it not only gives the participants the courage to dream beyond their means, but also it gives them complete ownership of their dreams. It was not a dream imposed upon them by some NGO or the government, it was their dream and it was up to them to make it come true.

Once the dream village was articulated, participants collectively thought of a statement that described the dreams they have for their *panchayat*. The statement reflected their dream as well as the values that the *panchayat* stood for. The vision statement of Dibburhalli GP is, "Namma Grama, Namma Nirnaya, Namma Abhirudhi" meaning "Our Village, Our Decision, Our Prosperity," signaling their independence in decision making and paving their path towards prosperity.

The second day of the workshop focused on the methodology used to obtain answers to three key questions for the Mission statement: WHO are the stakeholders? WHAT are their expectations? HOW can the services be improved? Essentially combining the grass-roots level methodology with the corporate methodology, the PRA exercises have been modelled in such a way as to elicit information on the above three questions. Five PRA exercises were to be used on the ground. Participants were trained on the PRA exercises, which were also demonstrated in a nearby village, for the participants to understand how to conduct them in their respective wards.

Diagnosis of the Current Status related to Key GP Functions

Process mapping tool was used for diagnosing the existing processes of the GP by breaking down its broad functions into detailed activities, helping the members to demystify their roles. Also, they articulated specific challenges *vis-à-vis* different activities.

The process mapping tool also tracked the RACI[4] *vis-à-vis* each activity, which helped the members understand the support they could rightfully demand from different agencies and that they alone were not responsible for delivery to the citizens. Lastly, the process mapping exercise led to decisions on how to improve the processes, identifying indicators which can be used to measure the successful implementation of these processes. This was a participative and enriching process. In addition to gaining awareness on current state, it was empowering for members to develop solutions against identified challenges, providing energy to make the system more effective rather than despondency with the present state of affairs.

Care was taken to ensure that processes being mapped reflect the priorities of GP members. We mapped 17 processes in the GPs, providing water through piped water supply, cleanliness of roads and drains, procurement, etc., which were derived from the vision and mission exercise and further discussions with the members.

Redesigning GP Organisation Structure

The GP structuring exercise began with collection of information on as-is structure, roles, budgets, and functions within the GP from various staff and members. Simultaneously, the as-is processes were finalised and consolidated to understand the current roles and responsibilities of GP members *vis-à-vis* the processes that were mapped, and the issues they faced.

Some key observations with this current structure were:

- Although line departments exist, citizens hold ward members accountable for most functions. Role boundaries between the GP members and the line departments are unclear.
- Liaison with line department is sub-optimal, due to which knowledge of their financial and manpower resources are unclear, leading to poor performance of programmes and schemes.

4. Describing roles as 'responsibility (R), accountability (A), collaborating (C) and providing Information (I).'

- Accountabilities are unclear, thereby impacting the performance.
- The GP *adhyaksha*, irrespective of capacity, has to single handedly perform many roles.
- Capacity building is primarily top down, due to which many development needs of a GP do not form a part of the curriculum by SIRDs and other training institutes.

A study of the skill levels in the project GPs revealed that there are broadly three types of GP members. Majority, about 50-70 per cent of the members are elected first time, are capable and motivated, but have limited knowledge about their role, government structure or programmes; 35-50 per cent of GP members are also elected first time, but are passive, for various reasons. A miniscule 10-15 per cent of the GP members are well-versed with their role and government structure, are capable and motivated, and are able to handle citizen pressure. Mostly the last category comprises of re-elected members. Key observations related to skills and knowledge within the GP included:

- Only a limited subset of GP members has both the capacity and motivation to perform organisational roles.
- Government staff is not sufficient. PDO, supposed to be earmarked for a GP, gets assigned to two or more GPs. GP paid staff need more skills. Also, they are over-burdened. For example, one bill collector may not be adequate to collect taxes from the entire GP.
- The standing committees, which is the only concept which aims to decide roles within a GP, are either not formed or exist on paper. Mostly, they are not functional in the manner envisaged by the KPR Act, 1993.
- Lastly, a point which we have raised earlier—despite assuming executive responsibilities, a GP member is not adequately compensated.

The GP members also articulated the strengths of the present structure, which are listed below:

- Governance structure has been decentralised as per the Act and powers have been given to GPs.
- All decisions have to be vetted by the *gram sabha*.
- *Adhyaksha* is elected by all villages in the democratic process.
- People can access GP for basic facilities.
- Resource mobilisation at local level is possible.

Proposed Organisational Structure

The proposed restructuring was designed to bring role clarity while aligning the GP with the structure of government line departments on one hand, and to utilise the potential of GP members and citizens, seeing them as resources and assets rather than as individuals constantly seeking more government functionaries to perform various functions, on the other.

- The structure was envisaged keeping the strengths in mind, as well as the legislative requirement of having Standing committees.

A brief description of the roles in the structure is listed below:

- *Head-Functions*: Accountable for service delivery of functions assigned to him/her (including pre-planning, planning and implementation), while working with GP members and others in the respective standing committee. This role will include liaison with respective line departments for better service delivery.
- *Head-Capacity Building*: Identifying training needs of other heads, members and staff and liasing with different training institutes for training. He/ she will also help source and build capacity of local people (based on database of citizen profiling) for small projects to be implemented by the GP.

Figure 14.3

Proposed Organisational Structure

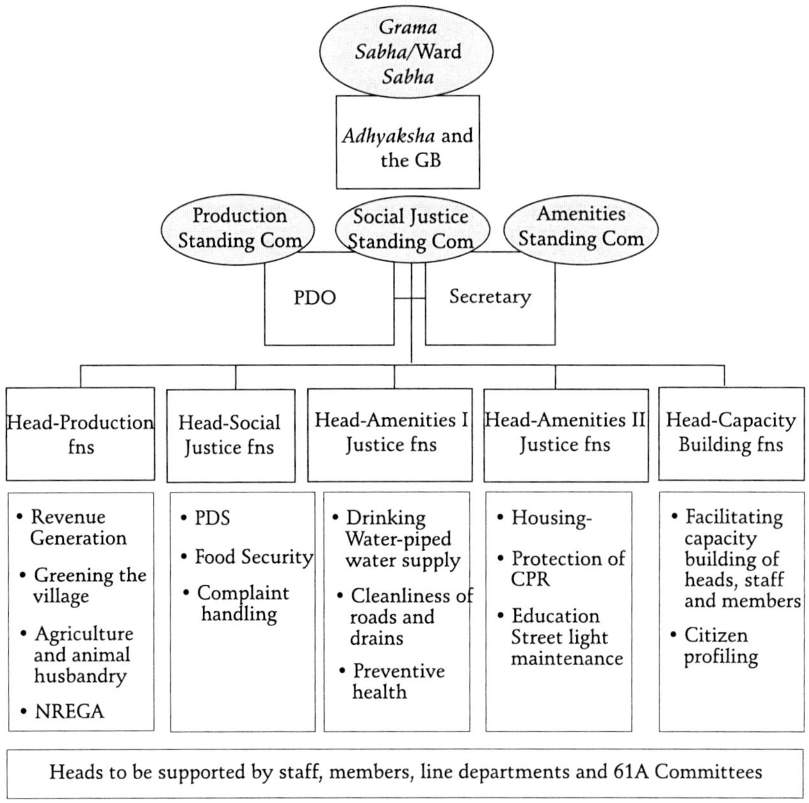

- *Role of Standing Committee*: Recommenders, but not decision makers for functions under the committee.
- *Role of GP body*: Approval of all plans and key decisions, subject to agreement of the *gram sabha* (no change).
- *Role of Gram Sabha*: As per the KPR Act.
- *Role of PDO/Secretary*: will help design and sign contractual agreements between line departments and the GP.

Benefits Observed

As the recommendations were being implemented in the two project GPs, we observed positive changes, some of which are listed below:

- Increased Revenue generation by GPs:
  - Water tariff—increase in percentage of collections
  - House tax—increase in percentage of collections
  - Exploration of new avenues—e.g., auctioning of eucalyptus plants, tax on mobile towers, auction of public places and parking lots, etc.
- Improvement in delivery of basic services by GPs. Examples of some service delivery improvements are mentioned here:
  - Drinking Water
  - Greening the Village
  - Public Distribution System (PDS)
  - Reduction in discrepancy between stock and requirements based on number of BPL families sanitation.
- Strengthening organisation capacity of GPs as well as ownership among elected members.
- Improvement in programme delivery and leveraging government resources, such as MGNREGA, National Rural Health Mission (NRHM), Sarva Shiksha Abhiyan (SSA), etc.
- Enhanced citizen participation and response time.
- Activation of standing and GP sub-committees.

Overall Recommendations

Based on our work with the two GPs, our key recommendations for building a strong GP revolve around the necessity and methods to develop a strong GP organisation:

- We need to invest intensively in strengthening the GP organisation, as a well-structured and organised GP will be able to understand and perform its own functions, while simultaneously ensure deliverables expected from other agencies such as the government line departments, hence triggering change at other levels too.

In addition, a strong GP can become the much needed convergence body for many government programmes, such as SSA, NRHM, etc. Though not immediately, the parallel structures created by these programs, such as School Development and Monitoring Committees (SDMC), ASHA workers, etc., can work under the GP, thereby leveraging existing resources better.

- OD methodology can be used to develop a framework for strengthening the GP organisation. The rationale for the using OD framework is its focus on organisation as a whole, and not on standalone components. The premise is that successful change occurs when all components of an organisation are aligned, i.e. the vision, structure and skills, incentives, resources, action plan and results. The GPOD project has evolved a detailed step-by-step framework, designed to create ownership for change among elected representatives (ERs), develop an enabling environment for ERs to function and ensure continuous improvements to sustain momentum.
- The key components of the GPOD framework include evolving the GP vision and mission; developing an organisational structure, elaborating various roles and responsibilities, to help the GP deliver the three mandates assigned to it by the 73rd Amendment, *viz.*, participation, improved service delivery; and functioning as a unit of self-governance. And finally, laying down systems and processes which will aid implementation and decision making in all key areas: participation, service delivery and governance. The annual plan of the GP should be derived from its vision, which reflects their aspirations, and is not a sum of government programmes and schemes.
- ERs are best placed to take on the functional roles in the *panchayats*. Even while focusing on service delivery, GPs need to tackle complex issues such as demand management with citizens, skills gaps among staff and funds shortage, etc. In fact, effective service delivery requires the GP to

address more complex social issues such as resolving caste disputes, ensuring gender equity, etc.

To be able to deliver these functions ably, there is need to rally citizens around key decisions. The government-appointed employee, such as the PDO, may struggle to understand local issues and is, anyway, not always accessible.

At 2.8 million, India has the highest number of ERs in the lowest tier of government.

- In its capacity as a local self-government, the GP should be able to sign service MoUs, stating mutual deliverables with government line departments and charge service fee for delivering services and pay compensation to its ERs who perform functional roles. Members and employees should be remunerated for their efforts. However, the recommendation is not to pay salaries from government funds, but for GP to draw money from its revenues to pay its members. Rather than creating a culture of entitlement, this approach will enhance self-sufficiency and, therefore, self-governance.
- Finally, raise local capacity to address local issues. Existing skill building and training institutes need to be leveraged. However, programmes and participation need to be decided by the GPs to raise capacity of its members, staff and citizens.
- In the absence of the last two recommendations, GPs will lack teeth and will be unable to motivate its members to perform in a sustainable manner.

Conclusion

The role of the GP in ensuring last mile delivery and governance cannot be undermined. The question is not whether we should invest in the GP or not. The focus needs to shift to 'how to make the GP stronger.' Different GPs may have different needs. There is need to evaluate the needs and focus our efforts accordingly. In our conversation with many experts, one of the all-pervasive gaps seemed to be lack of implementation ability in most GPs, which is closely linked to its organisation ability.

To build its implementation ability, there is need to focus on the GP organisation as a whole, hence the OD framework. The motivation and capabilities found in the ERs need to be channelised, to bring about true change and grass-roots democracy. Lastly, we need to create an enabling policy environment, which may include formalising agreements of the GP with other agencies, GPs charging service fee, paying compensation to their ERs and a training system which is demand driven.

15

JOE MADIATH and ANUSHA BHARADWAJ

Social and Gender Equality in Water and Sanitation

Introduction

Sixty-five years after independence, India is still struggling to reach out to rural India. Insular pockets of growth have further isolated villages from benefiting from the fruits of mainstream development programmes. Accessibility and availability of clean drinking water is the main concern of many Indians living in rural India. Dusky women carrying pots of water on their head, in the background of flute music, has been the portrayal of typical village scenery in many Indian films. But the romanticism ends with movies. Fetching water is drudgery for a woman. She is doomed to spend many an hour, scoring hills and plains, to fetch that precious pot of water for her family. Lack of clean drinking water, poor sanitation and poverty further send the poor into a spiral of rural indebtedness.

Gram Vikas is a rural development organisation working with poor and marginalised communities of Odisha since 1979, and presently, it is operational in 24 districts in Odisha. Since 1992, Gram Vikas has implemented its community-based sanitation and water supply programme, in which each family in the village builds its own toilet and bathing room, with piped water supply from a common overhead water tank, ensuring 24x7 access to protected piped water supply to all families all through the year. MANTRA (Movement and Action Network for Transformation in Rural Areas) is the overarching framework of Gram Vikas' habitat development initiatives. As of March 2012, Gram Vikas has provided 24-hour potable, piped water supply and sanitation facilities to 988 villages (encompassing 57,793 families).

The Context

Access to safe water is a basic human right. This has direct bearing on health and quality of life of the community. "The combination of safe drinking water and hygienic sanitation facilities is a precondition for health and for success in the fight against poverty, hunger, child deaths and gender inequality" (WHO, 2004). At least 1.1 billion people lack access to safe water, and 2.6 billion lack access to basic sanitation, a silent humanitarian crisis that each day takes thousands of lives, robs the poor of their health, thwarts progress toward gender equality, and hamstrings economic development, particularly in Africa and Asia (WHO/UNICEF JMP, 2000). Women continue to be the silent sufferers—carrying on with the burden of hauling water over long distances or facing the risk and indignity of having to defecate in the open.

Safe Drinking Water Crisis

India has made progress in the supply of safe water to its people, but gross disparity in coverage exists across the country. Although access to drinking water has improved, the World Bank (WB) estimates that 21 per cent of communicable diseases in India are related to unsafe water. In India, diarrhea alone causes more than 1,600 deaths daily. Hygiene practices also continue to be a problem in India. Toilet usage is extremely poor in rural areas of the country which is only 14 per cent. The Government of India's (GoI) Department of Drinking Water Supply statistics show 30 million people on average in rural areas of India do not have access to sanitation. This is compounded by the fact that sanitation being a lesser need felt by people; it hardly is an area where the household would like to invest upon.

The lack of proper sanitation is especially a serious concern for women who continue to face the indignity of having to relieve themselves out in the open in the absence of private latrines. In the search for discrete locations, women rise before dawn or wait after dusk to defecate. In the absence of a protected enclosure where women could take a bath, they are forced to bathe in the common village pond. In these ponds, due to the presence of men on the other side, women are mostly forced to resort to only insufficient cleaning of their bodies. Moreover, in summers, communal bathing in turbid waters of a shallow pond is an instant recipe for the spread of skin diseases.

Deep-rooted Exclusion

Exclusion is the bane of rural Indian society. Exclusion on the basis of caste, class, gender, etc., is a common practice in communities. The most vulnerable and marginalised face maximum exclusion. For example, woman is already treated as a secondary citizen in the community. If the woman happens to be a widow (even worse if she is from a lower caste), she faces exclusion and in most cases, is left out of most community decisions. These practices are deep rooted to the point where the excluded consider it their fate and those excluding them believe it is their right to do so. A successful water, sanitation and hygiene (WASH) programme should involve every family in the village without exception. This is essential to ensure that sanitation is effective and there is no open defecation or contamination of water bodies by any family. More importantly, 100 per cent consensus ensures that even the poorest and most marginalised benefit a minimum level of services, and has an equal vote in deciding how the project should be implemented.

Dual Challenges in WASH Sector

In WASH, there is a need to understand and integrate sanitation with safe water. In India, though both are under the purview of the same department, i.e., the Department of Drinking Water and Sanitation, the focus is more on coverage of water while sanitation continues to be the poor cousin. The link between sanitation and safe water is most often not understood by the implementers, who continue to adopt a disjointed approach, thereby becoming a mutually defeating venture.

The link between safe water and sanitation is obvious. Over 80 per cent of the contamination of water sources is through faecal contamination, which is a result of unhygienic waste disposal behaviours of the rural communities. Further, sanitation has suffered from a target approach, assuming people with economic capacities would be inclined to invest in the setting up their own infrastructure, whereas the poor people needed some assistance from the government. This has not proved effective, since it never achieved the stated aspiration 'Total sanitation.' The fact that monetary conditions do not necessarily constitute a factor for investment in sanitation is well proved.

An individual approach to sanitation may lead to increase in numeric coverage of toilets but is not likely to achieve safe water or better health. The families left out continue to pollute the water through open defecation, thereby reducing the effectiveness of the intervention. Thus total coverage and total transformation of waste disposal habit should be the approach in order to have improved impact. We will explore these two components in this paper.

Linking Sanitation and Health

To establish this link with the reality in Odisha, Gram Vikas conducted a study in 1992. The study showed that over 80 per cent of morbidity and mortality in rural Odisha was due to poor quality of drinking water, caused by the callous attitude towards human waste disposal. Another survey conducted by Gram Vikas in 2004 of 4,399 households in 49 villages across nine districts of Odisha, showed that less than 1 per cent had access to piped water supply.

Gram Vikas works in rural Odisha in predominantly tribal pockets which have very poor infrastructure development. Most of the tribal habitations have low population density and are located in hard to reach areas where there is lack of basic facilities like drinking water. Majority of the population depend on unsafe water sources—streams, springs and open wells. In summer season, villages face acute water shortage.

Open defecation is the norm in rural Odisha and one of the main reasons for pollution of water sources and waterborne diseases like diarrhea, dysentery, typhoid, cholera, etc. It is estimated that, every year, diarrhea alone kills more than 2 lakhs children under the age of 5 in India. Poor personal hygiene practices also contribute to large-scale morbidity that lead to people falling sick and losing critical working days.

While the government is mandated to provide drinking water to the citizens and there are efforts that seem to be there, sanitation is an issue that hardly gets much priority. Without sanitation, safe water continues to remain a distant dream. The target approach adopted by the government in case of sanitation without much emphasis on community participation, ownership and behaviour change is also a factor for low coverage of sanitation. Another reason for low usage of latrines promoted by the government under Total Sanitation Campaign is the poor quality infrastructure and lack of water facility for toilet use. On the ground, the water supply projects are also marred by dysfunctional systems, source failure, etc., and one of the primary reasons for this problem is the lack of community involvement in the planning, implementation and post-implementation operations and management.

Social Inclusion

The experience in Odisha showed that the practice of exclusion—of dalits, adivasis, lower castes, widows and women in general—has been deep rooted in the social fabric of society and it requires a great deal of conscious and courageous efforts to challenge the *status quo*. However, the journey to establish open defecation-free villages will require 100 per cent participation of the community and it necessitates a complete overhaul of the outlook of the community towards exclusion. Centuries-old practices must be broken and a new paradigm of development must be established. Democratic public spaces are to be widened and redefined where the excluded have a voice and the women are partners in this new framework.

The MANTRA Programme

Gram Vikas' MANTRA programme unites communities to overcome barriers like social exclusion for provision of drinking water. The programme addresses both improved health and transformation of hierarchical caste- and gender-based exclusion into equitable inclusion. Thus, the objective of Gram Vikas' work is to improve the quality of life of rural communities by improving their access and availability of drinking water thus enabling them to lead a dignified healthy life.

Specific objectives:

- To enable rural communities with improved access to safe drinking water facilities and proper sanitation.
- To enable communities to have improved hygiene and sanitation practices with respect to water use and waste disposal.
- To eliminate existing social and cultural discriminations in accessing water.

In MANTRA, every household in the village constructs for itself a toilet and bathing room, with 24-hour piped water supply to both the toilet and bathing room as well as to the kitchen of the house. The water is supplied from an overhead water tank constructed on the basis of estimates of per capita consumption of water (of 40 liters per day), projected for a population 20 years hence. This is done so that every household can have 24-hour running water supply on their taps. The water is pumped up using electricity where available and in the hilly areas, water is sourced from perennial springs using the gravity flow system. Where either of these possibilities does not exist, solar power has also been used.

As of March 2012, Gram Vikas has successfully implemented its flagship programme in 988 villages. The beneficiary communities are diverse in their socioeconomic-cultural and geographic backgrounds [i.e., scheduled tribes (STs), scheduled caste (SC) communities, general caste people, below poverty line families, people in hilly and undulated terrains, plain regions, etc.] This has been possible as the programme has strong core values. These non-negotiable core values of MANTRA programme are the cornerstones of every development activity undertaken. These values are:

- *Inclusion*: 100 per cent of the community must benefit from the programme, regardless of caste, social or economic status.

- *Equity*: Social equity and gender equity in all self-governance activities.
- *Sustainability*: Built in mechanisms for financial, social, environmental and institutional sustainability
- *Cost sharing*: The community contributes in cash or in kind (construction materials) and also renders their physical labour This enables them to feel ownership of the assets created by them with Gram Vikas, the government and other donor agencies.

Methodology

The implementation of a socially inclusive and democratic model of Gram Vikas' water and sanitation programme involves the following stages:

- *Motivation period:* When Gram Vikas enters into a partnership with a village, it requires the unanimous support and participation of all adult members of a community as a non-negotiable condition of the MANTRA programme. All households must be involved at all stages in the development process and must benefit equitably. This is based on the principles of inclusion and social and gender equity. For most part, women and the poorest families have been excluded from decision-making in the past. Hundred per cent consensus is also necessary because clean water cannot be assured until there is 100 per cent proper disposal of human waste.
- *Democracy*: Gram Vikas assists each village to set up a village general body, represented by every family in the village and members comprising of each head of household, male and female. In addition, a village executive committee (VEC) is formed and registered under the Society's Registration Act 1860; typically this will comprise of five men and five women from the village plus a Gram Vikas representative. The VEC has proportional representation of all the social classes and castes living in the village. This gives everyone a voice in decisions about village matters. The village

members collectively devise a plan for raising the long-term operation and maintenance costs (pump operator's wages, the electricity bill for pumping and the ongoing maintenance). Each household will contribute an average of ₹ 30 per month. Of late, most villages have started installing water meters to each family water connection and families pay volumetrically for the amount of water they consume, leading to greater equity. In addition the villagers take up community income-generating activities such as fish farming, cultivation of common land or giving a proportion of the harvest (with ¼ to ½% of the yield being contributed). Once agreement has been reached, before anything is commenced, the villagers collectively agree and sign an Memorandum of Understanding (MoU) between them and Gram Vikas.

- *Community fund:* The acid test of the community's resolve is the mandatory contribution of ₹ 1,000, on an average, per family into a community fund. In practice, households living below the poverty line contribute less, whilst better-off households contribute more. This fund is placed in a fixed deposit and the interest is used to cover the social cost (cost of external materials) of extending water and sanitation facilities as the number of households in the village increases.
- *Materials collection:* For the construction of the toilets and bathing rooms, all households are required to contribute in kind, providing locally sourced materials, including sand, aggregate and bricks (hand made by the villagers). The cost of externally sourced materials (pipes, taps, porcelain pans, cement, steel, doors and roofing) would be covered by the subsidy provided by funder/government/Gram Vikas.
- *Training people in masonry and plumbing*: Young men and women who are daily labourers can take a 75-day training course in basic masonry skills. Once trained under the technical supervision of Gram Vikas, they build the toilets, shower rooms and the elevated water reservoir of the village. They can have guaranteed work with Gram Vikas

for a period of two years if they so desire. Installation of the water supply and taps is also carried out by young men and women under the supervision of trained plumbers.

- *Health education*: There is no point in building toilets and bathing facilities unless they lead to a positive behavioural change in the sanitation practices of the community. The women and children in particular are informed about the importance and benefits of personal hygiene (including washing hands and trimming finger nails), the need to keep their new facilities clean and the necessity for all members of the family to actually use the facilities at all times. A sanitation committee inspects all toilets and bathing rooms regularly to ensure that the facilities are being kept clean. Fines are imposed for non-compliance as well as on people not using their facilities.

Sustainability

To ensure success of the programme, mechanism for sustainability must be stressed at every stage of the programme. Gram Vikas' programme incorporates mechanisms for sustainability for the social capital that is created:

Institutional sustainability: The villagers are capacitated to manage their own affairs through village committees.

Social sustainability: Mechanisms for inclusion are institutionalised to ensure equity and justice and thereby enabling communities to overcome barriers of caste, gender, etc.

Financial sustainability: Families contribute an average of ₹ 1,000, where the poor give less and the better-off give more as decided by communities, towards a village corpus fund which is invested and the interest earned is used to construct new toilet and bathing rooms for new families that may come up in the village after the project is completed. A maintenance fund is raised to cover operational expenses where villagers pay for the usage of water volumetrically (water meters).

Environmental sustainability: Gram Vikas does not mine water and wherever possible gravity flow from springs, induced gravity flow water supply systems or sanitary dug wells are built by the villagers themselves. This also ensures that there is no outflow of money from the community as dug wells can be constructed by the people themselves. Gram Vikas also undertakes appropriate water recharging and water harvesting mechanisms to recharge the aquifer so that the water sources do not run dry during summer.

Impact of the Programme

Studies have shown 85 per cent reduction in the incidence of waterborne diseases in these villages. In all these villages, over 90 per cent eligible children are in school and regular immunisation camps are conducted, covering all eligible children. The programme has had a positive impact on the quality of life of all participating villages, through reduction of waterborne diseases and, hence, a marked improvement in the health situation. Due to the importance given to personal and environmental hygiene, people have cleaner habits and are more aware of their responsibility in keeping not only themselves, but also their village clean. In all villages where water supply has been established, the communities have ensured that the systems are functional. In areas of women's empowerment, participation in *panchayats* or even in some cases, of inter-caste relations, the achievements till date have been heartening.

Conclusion

A development chain starts with improvement in health. It leads to better productivity, more income, access to education and ultimately a better quality of life. The Gram Vikas model has all the elements for success. Firstly, it involves the entire community and brings them together for their own development. Secondly, several mechanisms are put in place for creation of vibrant local institutions. Thirdly, the focus of the MANTRA strategy is dignity of women by reduction of drudgery of women and provision of toilets and bathing rooms. Fourthly, mechanisms for sustainability of the programme and clear withdrawal strategies are in place for every village. Lastly, it is a highly replicable model. Keeping the non-negotiable values intact, the model can be adapted to suit communities in rural India. In conclusion, Gram Vikas demonstrates each day that the poor deserve quality solutions and it is the role of development agencies to help them realise their potential. Note that a study conducted by PricewaterhouseCoopers (PwC) in 2007 on 'Gram Vikas NGO Network Opportunity to Grow into the Next Stage' states that 'Over the past decade and a half, Gram Vikas has organised a successful, community managed habitat development in the rural areas of Odisha'. The report continues to say that ('Gram Vikas') Rural Health and Environment Programme (RHEP) has proven success in Odisha and based on 15 years of experience of working with the concept, all major flaws have been solved and can be used outside. MANTRA is a strong and proven concept worthy of up scaling. The short-term focus should be on developing the Odisha network, continue outside Odisha pilots and build the network with other states when all the key recruits are in place and the experiences of networking can be leveraged. A matter of pride for Gram Vikas is that the Rural Development Ministry of the GoI has recognised Gram Vikas' MANTRA model as a sustainable model for wider replication throughout the country.

References

Gram Vikas' internal documentation on MANTRA. *http://gramvikas.org/index.php?act_id=2&page_id=11#mantra*

Jayapadma, R.V. and Liby T. Johnson (2003). "Institutionalising Gender in Patriarchal Rural Communities: Creating Spaces through Uncontested Domains, Genre", *Pouvoirs et justice social le, Cahiers genre et development no. 4.*

Joe, Madiath and R.V. Jayapadma (2002). *A Common Cause—NGOs and Civil Society—Strengthening India's Villages the Sustainable Way: A Common Cause*. New Delhi: National Foundation for India.

WHO (2004). *Meeting the MDG Drinking Water & Sanitation Target: A Mid-Term Assessment of Progress*. UNICEF and WHO.

WHO/UNICEF JMP (2000). *Global Water Supply and Sanitation Assessment 2000 Report*. WHO and UNICEF.

16

YUSUF KABIR, SHASHANK DESHPANDE and
AJIT PHADNIS

Governing for Sustainable Water Service Delivery at Village Level

Experience from Maharashtra

Introduction

Ever since independence, India had invested an enormous sum of nearly ₹ 68,000 crore (over $13 billion) towards improving water supplies in rural areas through various schemes and programmes during the plan periods till 2007 (IRAP, UNICEF and WSSD, Government of Maharashtra, 2013). The aim underlying these efforts was improving the coverage of the schemes in terms of number of villages and hamlets.

Although remarkable achievements were made in terms of improving the coverage *vis-à-vis* the habitations and population, the overall achievements are not commensurate with the scale of investments that have gone into the sector. One important factor, among others, is the increasing number of habitations slipping back to the 'no-source category,' or no year-round adequate supply of water is available. This is due to several threats that rural water supply sector is facing and can be summarised as: poor sustainability of water supplies owing to unsustainable resource base; soaring operation and maintenance (O&M) costs due to absence of regular systematic upkeep; and poor cost recovery. This results in poor financial management and operation of the system and has led to inequity in access to water across different segments of society. There is a critical need to address the lack of adequate institutional capacity at the local level for water distribution.

Only half the rural households have access to tap water, and it includes those who access it from nearby sources and distant sources. Out of this, only 63 per cent of the tap water is treated. But, in terms of high quality water supplies, only 20.6 per cent of the rural households have access to treated tap water within their dwelling premise. If we include those households which get treated tap water near their dwelling premise, the percentage would go up to 29.5, i.e., another 12 lakh (1.2 million) households. Nearly 20 per cent of the rural households still depend on distant sources for their domestic water supplies. Of this, more than 45 per cent depend on unprotected sources including uncovered open wells and canals, springs, ponds, etc. Nearly one-sixth of the rural households (15.7%) depend on hand pumps for domestic water supply. But, only in 14 per cent of the cases, the households own these systems, and therefore have complete control over their functioning and management. In the remaining 86 per cent of the cases, the households are dependent on sources which are publicly owned, and hence have no control over the management of these systems [Census of India, Government of India (GoI), 2011a].

Management of water service delivery is a responsibility of decentralised institutions of governance i.e., *gram panchayat* (GP), *panchayat samitee* and *zila parishad* (ZP). GP is the lowest level of local governance institution closed to the communities. Their capacity to govern is the most vital parameter. The government flagship programmes are aligned with this perspective. It gives prime importance to the GP and village level processes.

Progress and Challenges

Maharashtra is one of the well-endowed states in India which receives very good rainfall mainly from the south-west monsoon. However, at the same time, there is tremendous variability in rainfall ranging from 400 to 6,000 mm. Besides rainfall, the overall availability of water in the state is dependent on geomorphology and geology.[1] The geomorphology and geology too are not favourable with rapid runoff and low storage capacities.

1. Groundwater surveys and development agency (Government of Maharashtra) and Central Groundwater Board, Government of India (2008-09).

Over the years, Maharashtra had tried a variety of rural water supply systems, with increasing focus on decentralised O&M. While some of them have been successful (Gawde, 2012), some have failed on several fronts. The reasons for failure were: technical systems for water supply often designed without due consideration to the local hydrological system or the socioeconomic realities; appropriate policies with regard to level of investment, staffing, water supply technology, coverage of systems, and O&M were not followed. Due consideration was not given to the capabilities of the local agencies in handling O&M of the scheme, and the poor did not have adequate incentive to access water from the public system, severely affecting their willingness to protect them from damage and to pay for these services.

The rural drinking water supply policy of Maharashtra has changed in keeping with the National Rural Drinking Water Programme (NRDWP) and external funding agencies. Initially, the emphasis was on village level schemes which changed to multi-village schemes, known as the Regional Rural Water Supply Scheme (RRWSS). The RRWSS were taken up in a big way from the late 1980s. These are fed from large dams in which 5 per cent water is reserved for drinking water purpose. The Government of Maharashtra (GoM) has been also pursuing their goal of 'no tanker-fed village' by supplying water through such large schemes; this approach was purely supply driven. Funds from the Department for International Development (DFID) of the British Government for the Maharashtra Water Supply and Sanitation Project helped to some extent in pursuing the goal of no tanker-fed village. The Maharashtra Jeevan Pradhikaran (MJP) was mandated for implementation of the RRWSS and the schemes, once installed, were to be handed over to the ZPs (locally elected government bodies) for maintenance. The ZPs, however, had much reluctance in taking over maintenance responsibilities of these schemes. There was a general disillusion on the operation of such large schemes by the end of the 1990s and the approach for having no tanker-fed village through RRWS was not very successful due to huge cost involved (Das, 2006). There were also problems like unequal distribution of water between tail-end and head-end villages and unwillingness of taking over of the schemes by the ZPs, mainly due to capacity gaps.

Around the late 1990s, a new approach, catalysed by a funding crunch as well as the general influence of neoliberal policy around the world in treating water as a commodity to be paid for instead of a public good to be supplied as a right, was adopted in the rural drinking water sector at the national level and the GoM also adopted the same. The other motivation behind the shift was, at least on paper, more decentralisation of functions to rural local self-government post the 73rd Amendment of the Constitution which pushed more powers to locally elected bodies. There were corresponding shifts in the processes and institutional responsibilities. Under the new policy, the GP, or village-elected body has to approach the ZP and the decision on the nature of scheme, the implementing body [GP, ZP or non-governmental organisation (NGO)] and the technical service providers was to be taken by the village committee or *gram sabha*. The Village Water and Sanitation Committee (VWSC) (Mumbai Gram Panchayat Act 1958), which is elected by the *gram sabha* and functions as a sub-committee of the GP, plays an important role in planning and implementation as well as mobilising 10 per cent local contribution and cost recovery. Thus, contrary to the earlier arrangement, the responsibility of the state or its agency in initiating water supply project was reduced.

The policy shifted to village-based schemes with emphasis on cost recovery of water being supplied as a commodity was also facilitated by the World Bank (WB) financed Maharashtra Rural Water Supply and Sanitation 'Jalswarajya' Project, which was implemented in the state from August 2003 to September 2009 with total project cost of $268.60 million (WB, 2012). The development objectives of this project were to: (i) increase rural households' access to improved and sustainable drinking water supply and sanitation services and (ii) institutionalise decentralisation of rural water supply and sanitation (RWSS) service delivery to rural local governments and communities.

In addition, in the last decade, the area under irrigation, and the associated water needed, has increased by 10-fold. Now the area under irrigation is over 4 million ha (Water Resources Department, GoM, 2012). and continually increasing, though the water resource body remains finite. Therefore, increasing emphasis on the development and implementation of new and more efficient approaches to water management is required. Similarly, the community must realise the importance of quality data around measurement of rainfall, availability of water, variability in the availability of average water, annual availability of groundwater, evaporation, etc., if it has the responsibility of their effective water management.

Decentralisation of Drinking Water Supply

The policy shift for taking up smaller and village-based schemes requires devolution of more responsibilities on the local government, i.e., to the GPs and to put in place necessary collateral measures so that the GPs are in a position to discharge their responsibilities effectively. This is in consonance with the 73rd Amendment of the Constitution which introduced the 11th Schedule listing subjects which may be devolved on the *panchayats* (Mumbai Gram Panchayat Act 1958). 'Drinking water' has been listed in the entry 11 in that schedule. A framework for implementation of NRDWP was issued in 2010, which provided that "The fundamental basis on which drinking water security can be ensured is the decentralised approach through local Government and community involvement." Assigning responsibility of drinking water supply on the local government, particularly the GP, puts in place a better accountability mechanism, promotes appropriate design of the system based on local context and wisdom of the local people, and facilitates taking ownership by the local people which helps better management and maintenance of the physical asset created as well as convergence with other activities and conjunctive use of water. Also, there is evidence that decentralisation of rural water supply (RWS) improves quality of services (Riddhi Foundation and UNICEF, 2013). All these arguments make a strong case for decentralisation of RWS (Ibid.).

However, decentralisation is not a panacea and to be effective, it has to be properly designed. Detailed Activity Mapping of functions and the roles and responsibilities involved and the corresponding requirement of funds and functionaries are to be carried out, and several collateral measures are needed to be put in place. Based on the Activity Map, specific functions are to be assigned to government and different tiers of local government with clearly defined responsibilities and devolving fund and functionaries. Most states have taken some initiatives in this regard, though such devolution is generally far from what is necessary for *panchayats* to deliver sustained drinking water supply to its residents. Model Activity Maps for functions, fund and functionaries of the NRDWP have been developed as an illustration by the Expert Committee on Improving Service Delivery by the *panchayats* headed by Mr Mani Shankar Aiyar (GoI, 2011a). The committee recommended that each state should carry out their own Activity Maps for each of the centrally sponsored schemes (CSS) to suit their local conditions.

In order to ensure sustainability of systems and water sources, appropriate cost recovery measures need to be incorporated. Allocation for O&M in the NRDWP increased from 10 per cent to 15 per cent of the allocation. At the time of transfer of the scheme to the Managing Committee/VWSC, an O&M Corpus fund amounting to 10-15 per cent of the project cost should be granted to the committee to supplement their income for O&M. A minimum collection of 50 per cent of O&M cost (including electricity charges) through user charges is advocated. Water metering, both bulk and individual household, should be promoted in all piped water supply schemes to reduce unaccounted for water (Ministry of Drinking Water and Sanitation, 2011).

Earmarked funds are also provided for the implementation of water supply to weaker section of society, such as scheduled castes (SCs), scheduled tribes (STs), other weaker sections of society, Integrated Action Plan (IAP) districts, Desert Development Programme (DDP) areas and north-eastern states. It is recommended that 10 per cent of the NRDWP allocation be given to IAP districts, i.e., 'rural population managing rural drinking water supply schemes' in the criteria for fund allocation to the states. The guidelines further clearly describes carrying out Activity Mapping by respective state governments to indicate the process, time frame and incremental improvement towards transfer of 'funds, functions and functionaries' to the three tiers of local government. Decentralisation in Maharashtra in rural water supply is arguably ahead of many other states in the country. However, further improvement is necessary in devolution through proper Activity Mapping of 'funds, functions and functionaries' related to drinking water supply along with putting in place several support measures like supportive supervision by ZPs and source strengthening for improving service delivery.

Change in approach demands change in institutional set-up. The MJP was strengthened to plan and implement the RRWSS in a centralised implementation mode. The current thrust on implementing small schemes by the GPs will require institutional rearrangement in which MJP will become a mere facilitator and technology provider for the GP that has the responsibility to install and maintain the schemes. That should require strengthening the engineering capacity for designing and implementing schemes, freedom and capacity to outsource services (from MJP or other providers in the market) in case the technology involved is beyond the competence of the GP and improved governance of water supply schemes. The latter will require augmentation of their capacities for improved O&M, sound accounting practices, and strengthening the VWSC for effective involvement and ownership of the community.

Maharashtra Water Supply Status

Coverage

Maharashtra state has made considerable progress for improving the coverage of rural water supply. The rural population of Maharashtra is 61.55 million and these are spread across 41,200 villages or 100,683 habitations. Table 16.1 presents the water supply status using the habitation (sub-village) as the unit.

As per the NRDWP programme, the habitations and population are categorised as fully covered, partially covered and quality affected. Fully covered implies that the water supply norm of 40 litres per capita per day (LPCD) is achieved, while partially covered implies the supply rate is less than 40 LPCD but more than or equal to 10 LPCD (Rural Development Department, GoM, 1999).

Table 16.1

Coverage of Water Supply in Rural Maharashtra

	Habitations			*Population*		
	Fully Covered	*Partially Covered*	*Quality Affected*	*Fully Covered*	*Partially Covered*	*Quality Affected*
Number	87448	11564	1671	498.13	117.30	35.34
Percentage	87	11	2	77	18	5

Source: (*http://indiawater.gov.in/imisreports/nrdwpmain.aspx*)

Maharashtra's water supply coverage is much better as compared to the country's average. In India, as on April 1, 2012, 74 per cent habitations are fully covered, 20 per cent are partially covered and 6 per cent are quality affected, whereas in Maharashtra, 87 per cent habitations are fully covered, 11 per cent are partially covered and 2 per cent are quality affected (NRDWP, 2012). This clearly shows that Maharashtra is continuously taking efforts not only in developing the sources and systems but also in trying to make them sustainable. This is attributed to the policy reforms of the state like decentralised water governance after the 73rd Amendment. The GoM has restructured the water supply sector and transferred a part of technical staff from MJP and Groundwater Survey and Development Agency (GSDA) the technical cadre to ZP working under CEO and collector.

Table 16.2

Progress of NRDWP in Maharashtra

Year	*Villages/Wadi Tackled*	*Expenditure*	
		Target	*Achievement*
2007-08	14,975	9,785	1,021.12
2008-09	19,877	19,877	943.64
2009-10	11,639	7,465	1033.63
2010-11	9,745	8,987	1,275.48
2011-12	6,502	6,364	1,252.92
2012-13	5,940	1,686	428.69

Source: Economic Survey 2012-13, GoM.

Although capital outlays are large, per capita expenditure under NRDWP comes only to approximately ₹ 200 (or less than $4). There is need for dovetailing of more funds, particularly that from Finance Commission awards, convergence with rural employment schemes and improving efficiency of the schemes to be taken up so that more people are benefited. This will require bottom-up decentralised planning and more community ownership for mobilising local resources including non-monetary resources.

Water Quality

The quality of water supplied for domestic purposes is of critical importance in terms of the achievement of public health results. The domestic water supply has to comply with water quality standards. There are two main types of potential challenges for water quality, bacteriological and chemical (inorganic and organic). Water quality in the state is monitored by the public health department which is reported under NRDWP and also by GSDA.

Chemical Contamination

As on April 1, 2012, there are 2 per cent habitations which have a problem of chemical contamination affecting 5 per cent population. The details of chemically contaminated habitations as on April 1, 2012 are given in Table 16.3.

Table 16.3

Extent of Contamination

Contaminant		*Number*	*Percentage*	
Fluoride	Habitations	483	29	
	Population	1178100	33	
Arsenic	Habitations	0	0	
	Population	0	0	
Iron	Habitations	337	20	
	Population	486161	14	
Salinity	Habitations	342	20	
	Population	602432	17	
Nitrate	Habitations	509	30	
	Population	1267072	36	
Total	Habitations	1671	100	2% total habitations
	Population	3533765	100	5% total population

Source: Numbers have been taken from NRDWP website, April 1, 2012.

Nitrate affected habitations are maximum; they are 30 per cent of the quality affected habitations, followed by fluoride 29 per cent, salinity 20 per cent and iron 20 per cent.

Bacteriological Contaminations

As per the water sample examination at public health laboratories in June 2013, 20 per cent of rural samples and 11 per cent of urban samples are bacteriologically contaminated.[2]

As per the rapid risk-based water quality profiling exercise of 15 districts, 186 blocks, over 1,200 habitations with 6,000 samples were taken up by UNICEF with GSDA and public health laboratories; the percentage of bacteriological contamination of tested water samples ranges from 34 per cent to 83 per cent (Report on M-DAWS, Maharashtra State, 2010), but overall 53 per cent of samples were found to be contaminated.

Table 16.4

Criteria for Issuing Different Colour Cards

Card Colour	*Criteria*
Red	GPs where: • water treatment is not regular • areas around water source are not clean • there are leakages • adequate stock of TCL powder is not available • there is outbreak of diseases during the year
Green	GPs where none of the above problems are reported
Yellow	GPs where areas around water source are dirty
Silver	GPs where no diseases outbreak has occurred during past three years

The bacteriological contamination is probably predominantly human induced. Insanitary conditions near source, inappropriate behaviour of water handling, faulty systems with lack of proper O&M are the most crucial factors leading to bacteriological contamination of sources as well as large amounts of faces in

2. State Public Health Laboratory (2013).

the environment due to open defecation; 62 per cent of rural Maharashtra do not use toilets. The public health department has initiated a programme to convey the villagers the status of their domestic water in terms of quality. Under the programme, villages are issued green, yellow and red cards based on the status of water quality, sanitation, operation and maintenance of water supply structures and outbreak of diseases.

Most practised and convenient way of disinfecting domestic water in the rural areas is chlorination. Bleaching powder ($CaOCl_2$) is used for disinfection. The chlorine content in the beaching power is also a challenging issue. Small things like proper storage of bleaching powder, administrating doses, monitoring of the residual chlorine along with proper behaviour change communication can significantly improve the situation. A system has been developed in the state to regularly test the chlorine content of the bleaching powder and report to the concerned authority. A sensitisation campaign has also been initiated to store the bleaching powder in air tight container in the villages. It has been observed with decrease chlorine content in the bleaching powder, the bacteriological contamination increases.

A Water Quality Protocol was developed by GoM with support from UNICEF to address the issues of regular water testing (both bacteriological and chemical) and its close monitoring at different levels, and it is currently displayed in all the village *panchayats* and public health centers. Displaying this protocol can be an effective tool of demand generation, helps to clarify accountability to citizens and is also an effective behaviour change communication tool also.

Groundwater

About 85 per cent of rural population is dependent on groundwater for domestic water supply. As per the *Report on Dynamic Ground Water Resources of Maharashtra* (2008-09) GSDA (GoM) and Central Groundwater Board, GoI, 2010) about 94 per cent of the total groundwater draft is being used to irrigate 69 per cent of the total irrigated area in Maharashtra. About 27 per cent of the total groundwater is consumed for irrigating 4 per cent of sugarcane area and only 36 per cent of groundwater is consumed for irrigation of 51 per cent food crops.

The management of ground water at decentralised level is a crucial challenge. Both demand and supply side managements are vital for source sustainability. This is especially important for the scarcity affected area. Around one-third of Maharashtra is drought affected (Report on Water and Irrigation Commission, 1999).

Division-wise status in terms of expenditure made and number of tankers in operation during the year 2012-13 (as on 4th March) is shown in Table 16.5.

Table 16.5

Tankers in Operation and Expenditure

Division	*No of Affected Villages/Wadi*	*Tankers in Operation*	*Expenditure (crore)*
Pune	550	3,197	760
Nashik	383	1,134	433
Aurangabad	771	378	999
Amravati	75	-	88
Total	1779	4709	2280

Source: Economic Survey 2012-13, Government of Maharashtra.

It has been estimated that more than 11,800 villages in 14 districts of Maharashtra are facing an acute scarcity of drinking water this year. Also, the number of drought-prone blocks has increased from 12 in 1960 (Fact Finding Committee report), to 17 in 1973 (Sukhtankar Committee) and ultimately to 26 in 1987 (Subramaniyan Committee) in spite of making a huge expenditure for improving drinking water supply in the state (Dandekar and Naravade, 2013).

O&M of Rural Water Systems

It is the responsibility of GP and VWSC to operate and maintain the schemes. For regional multi-village schemes, the responsibility lies with ZP and in some cases, with multi-village water sanitation committees.

Table 16.6

Functional Status of Various Water Sources in the Region

	Total	*Percentage of Defunct*
Hand Pump	1,83,899	14
Open Well	81,326	13
Mini W/S	13,466	12
Single Village PWS	40,261	15
Rural Regional PWSS	673	24

Financial sustainability is of crucial importance. The tariff structure and the recovery should be such that it would make the schemes self-supporting. The bottleneck analysis conducted by UNICEF in two districts of Maharashtra (Latur and Chandrapur) suggests that the recovery rate is below 50 per cent (*Bottleneck Report*, UNICEF, 2012). Technical procedures for both operations and maintenance and repair are also important. Capable institutions at village level can only be potentially deal with both these issues. Most of issues in RWS are expected to be dealt at village level.

Slip Back Habitations

Slip back habitations are those which fall back to partially covered status from fully covered status for various reasons. As per NRDWP data in 2012-13, 4,365 habitations have slipped back, just over 4 per cent of the total. Out of these, about 37 per cent are slipped back due to drying of sources. Jalna, Kolhapur and Buldana districts have reported more than 10 per cent of slipped back habitations. The highest number of slipped back habitations due to drying of sources is from Jalna district. The *talukas* (subdistrict level unit) where there were maximum numbers of slipped back habitations due to drying of sources were Jath from Sangli district, Solapur South, Ashti and Deoli from Wardha, Akot and Telhara from Akola, Soegaon from Aurangabad, Deoni, Jalkot, Shirur Anantpal and Udgir from Latur district. Most of these *talukas* are drought prone yet irrigated.

Village Water Account Initiative

In 2003, for the first time in Maharashtra, the government has declared water scarcity in the month of September in 69 *talukas* of 11 districts. However, after detailed study and analysis of actual rainfall in these 69 *talukas*, it has been observed that 4 *talukas* received more than 100 per cent rainfall, 8 *talukas* received rainfall between 80 and 100 per cent and rest 55 talukas had received less than 80 per cent of rainfall. Even though one-sixth of these *talukas* have received average/more than average rainfall, the crop situation was poor due to regional variation in the rainfall. Hence, GSDA initiated the village level watershed water account activity (Deshpande Shashank, 2009) on a pilot basis throughout the state with support from UNICEF, Mumbai office.

To facilitate management of water at village level and support their efforts to become self-sufficient for water, it is essential for these villages to know how much water is available to the village from the rainfall. Other questions will include what is the drinking water requirement (along with animals) and how can it be fulfilled? What is the present irrigation requirement as per the existing cropping pattern in the village and how it can be justified with the actual use of water? All these questions need to be answered through the village water account. Hence, the GSDA started the activity with people's participation in 349 villages of the state in June 2004. The motto behind this activity is to revive the traditional village water management system and to make the village self-sufficient for all uses of water.

Execution Strategy

Experts in hydrology, hydrogeology, agriculture, water supply, etc., were consulted and a standard strategy for implementation was developed. The standard execution strategy is as follows:

- Selection of pilot villages where there is community initiative and participation.
- In the selected villages, the following works are being carried out by the community with the technical assistance of *talathi* (revenue officer), *gram sevak* (village development

officer), school teacher, agricultural assistant, irrigation engineer and geologist.

a. Installation of rain gauge in the village.

b. Measurement of rainfall and groundwater level with the assistance of students or the interested/learned villagers/teacher (at least 3-4 persons).

c. Preparation of hydrograph, locally called as *jalalekh*, either on blackboard or on drawing sheets.

d. Collection of crop-wise data, latest census data and livestock population of the village, well census data, borewells with type of pump sets and pumping hours, etc., and existing water conservation structures with capacity. Numbers of big trees within the village are being listed separately.

e. Water requirements and their use for industries, hotels etc., if any, are worked out separately.

- Geologists of GSDA supervise and monitor on a monthly basis the work of measurement of rainfall and groundwater level during rainy season and thereafter.
- Based on the data collected and compiled at the village level, the water requirement and use for different purposes like drinking, irrigation, industry is calculated and a complete water account is prepared by the geologist of GSDA with the involvement of other officers, based on the hydrometeorological data, the surface and groundwater availability, actual retention of soil moisture, water requirement for drinking, agriculture and irrigation, industry and actual use from different sources, evaporation, etc.
- This water account is then kept before the *gram sabha*, held during the month of October, wherein the implications along with the benefits are discussed.
- Based on the water availability and its utilisation, a village level strategic action plan is finalised and policy decisions

are taken based on the deficit or surplus water, within the *gram sabha* only. The *gram sabha* is committed to implement their decisions.

- Priority is given for the security of drinking water at village level and accordingly *gram sabha* takes the required decisions. The water account is prepared only for the yearly volume of water that is replenishable. Hence, when needed a ban on construction and use of irrigation borewell may be imposed through people's participation and *gram sabha* monitors this.
- If the water account is in deficit, then a conscious decision is taken up for increasing the measures of artificial groundwater recharge. It is proposed to use water at up to 70 per cent dependability with the aim to raise the water table level. Accordingly a tentative plan is prepared and cleared in the Gram sabha. This is then submitted to the ZP for further processing and implementation.

Transforming Village Water Accounting to Village Level Water Safety and Security (VWSS)

There was a need felt to transform the entire village water accounting and linking this to the overall objective of village water safety and security process and align this to the overall objective of NRDWP. The idea behind aligning this with village water security plan is to concentrate more on drinking water safety and security and put across this as a priority to the PRIs. The skill set with the PRIs in visualising and implementing village water safety and security (VWSS) is very limited and without supportive supervision, conceptualising this idea is difficult. While doing so, the ultimate goal would be to institutionalise the VWSS into the routine system of NRDWP and with other state schemes.

UNICEF, in partnership with GSDA and technical services from PriMove, has been supporting Latur and Chandrapur districts from Maharashtra, India, to facilitate VWSS since the 2010. Initially the programme had local NGOs as partners and currently it has been integrated with the government flagship programmes under Nirmal

Bharat Abhiyan (NBA) and NRDWP. One of the main themes of the project is to ensure village level water safety and security, key concepts as per the NRDWP and international efforts on water management (e.g., WHO, WSP, etc).

Management for sustainable VWSS is a dynamic process. While one time capital investment in the infrastructure is necessary, it is inadequate to ensure the sustainability of VWSS. Sustainable measures involve water safety in terms of management of safe water quality, water security in terms of source reliability to deliver adequate quantity of water, and water systems in terms of financial, technical and institutional sustainability. Proper environmental sanitation and hygienic behaviour at the community and household level are the key determinants of safe water. Similarly, community ownership and their ability, willingness to pay and maintain systems are also vital from the sustainability point of view.

Components of VWSS Plan

The planning process involves:

- Establishing meaningful partnerships.
- Triggering.
- Demonstration.
- Participatory assessment and analysis.
- Technical inputs before and during micro planning.
- Preparation of action plan with clearly agreed ownerships and timeline to be used for implementation monitoring.
- Approval by *gram sabha* (a body consisting of persons registered in the electoral rolls relating to a village comprised with the area of *panchayat* at the village level).
- The action plans are presented in the form of a work book which has a basic framework and a space to reflect upon the issues which may be unique to the status of a particular village.
- Sharing with district stakeholders and their ownership.

The village level planning process (Figure 16.1) are divided in three sub-sections.

Figure 16.1

VWSS Planning

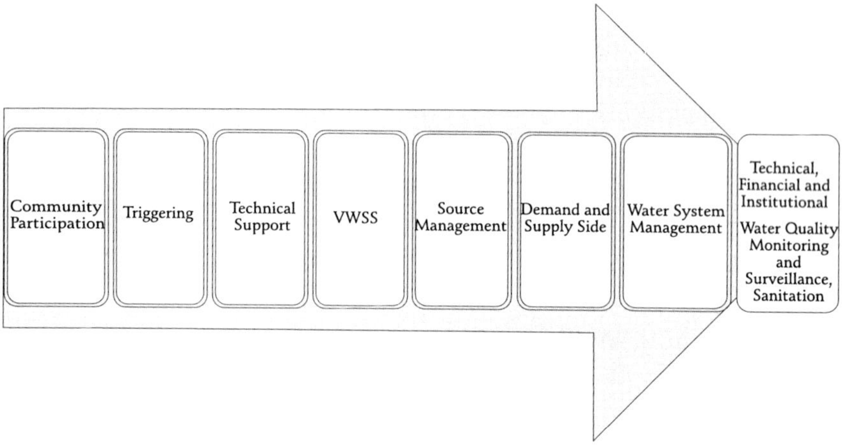

Water Security for Source Sustainability

- Measurement of rainfall and ground water level.
- Village water account.
- Planning and implementation to meet the water deficit priority basis.
- Facilitation of the demand side management actions such as cropping patterns, water use efficiency, etc.

Water Safety for Water Quality and Environmental Sanitation

- Regular water quality monitoring.
- Regular monitoring of residual chlorine, proper storage of chlorine and maintaining the ortho toluidine (OT) register.
- School sanitation action plan and hygiene promotion activities.
- Sanitation action plan at the household and village level including construction and usage of individual household level latrine (IHHL) and solid & liquid waste management (SLWM) activities such as cleaning of drainage lines,

construction of compost and soak pits and promotion of the biogas.

System Sustainability

This involves O&M of the full water system:

- Technical: Actions for repair, maintenance and monitoring of the performance of water system.
- Financial: Preparation of O&M budget, tariff setting and water tax collection system.
- Institutional: Institutional arrangement, roles and responsibilities, training of water person, accountability and reporting.

The preparation of VWSS plan requires significant capacity building at the various levels. Capacity building here needs to be understood as a 'process to strengthen the abilities of people, organisations and systems to make effective and efficient use of resources in order to achieve their own goals on a sustained basis.' The capacity building activities have been completed in total of 110 GPs spread across 11 community development blocks in the districts of Chandrapur and Latur where block resource coordinators, NGOs and VWSC members are trained on the concept and process of preparing VWSS plans, its implementation and O&M.

Result and Learning

The process has been completed in all the villages and the results are encouraging. The water tax recovery increased by 30 per cent points on an average from the base figure of 40 per cent, sanitation coverage increased and 50 per cent of the villages became open defecation free, all the GPs adopted different kinds of source strengthening measures, improvement in the storage of bleaching powder, regular chlorination, maintaining registers and all the GPs identified observations wells and installed rain gauges.

Mobilising the government resources through various programmes is a challenge. There are multiple government programmes which can provide for the resources required to implement village water safety security plan. Each of the government resource is tied up with a particular scheme, which has its own implementation arrangement and a process. Accessing those resources needs involvement of the officials of a particular scheme. For converging resources from these schemes, it is necessary to have a coordination at a higher level, i.e., block and district. If the entire process of planning is not owned by the officials, it is practically difficult to mobilise the resources.

If governing agency has an opportunity/mandate but does not have necessary ability, they can properly govern and a capable agency without a mandate also will not be able to govern.

The government programmes like NRDWP and NBA have created the opportunity for GPs to govern for improved water management. Adoption of VWSS process has enhanced the ability of local government agency to take informed decisions at the village level and to implement decided actions. Ability is a function of information, attitude, knowledge, skill and most importantly, systems. Enhancement of ability needs more focused interventions. VWSS process has contributed to enhancing the ability of GPs and communities. Annexure Table A-16.1 indicates the ability to do the process and also reflects on the results.

In the current phase, a VWSS intervention to enhance the ability is directly embedded in the flagship programmes and works with the mandated stakeholders. This is leading to improvement of the system. So far in two districts, activities such as capacity needs assessment, training of trainers, development of planning and development of activity-based implementation monitoring systems have been completed. In both the districts, village level planning has been completed in 215 out of 292 villages selected under one of the flagship programmes (NBA). The state level agencies are informed and the approach is being scaled up in other districts and incorporated in the Annual Implementation Plans (AIPs).

As mentioned earlier, both opportunity and ability are required for the effective governance. VWSS is a systemic intervention for process institutionalisation and capacity building, which is important for sustainable service delivery. Embedding approaches like VWSS in the mainstream government programmes, rather than stand-alone capacity building interventions, can enhance the chances of effective governance at the lowest level.

Conclusions

Maharashtra has adopted decentralised community-led demand-driven approach for ensuring safe and secured water supply in the rural areas. GP, *panchayat samittee* and ZP are mandated with the responsibility. They are supposed to operate and maintain the schemes through full cost recovery. Maharashtra has achieved about 85 per cent coverage for rural water supply.

Eighty-five per cent of the habitations depend on groundwater resource for their domestic water needs. Only about 6 per cent of the total groundwater draft is used for domestic water and about 94 per cent is used for the irrigation. The state has made considerable progress in terms of coverage, but the achievements are challenged mainly due to issues of sustainability of sources and systems.

The RWS policy architecture and the water resource base both indicate the governance of water service delivery has to be decentralised. Effective governance requires both opportunity and ability of the local government agency. GP, which is lowest level of governance agency closest to the communities, has been mandated by the national flagship programmes but ability needs to be enhanced.

VWSS approach enhances the ability of the GP and communities to ensure sustainable and safe water supply. The practical application of the VWSS approach in two districts of Maharashtra has improved the performance of GPs in not only planning and implementation of water supply interventions but also the O&M, indicated by better management practices like regular chlorination and increased tax recovery.

The government flagship programmes like NRDWP have addressed the sustainability issues in their programme framework through provision of resources and also adopting a decentralised approach.

Embedding systemic approaches like VWSS in the mainstream government programmes, rather than stand-alone capacity building interventions, can enhance the chances of effective governance at the lowest level.

References

Dandekar, A. and S. Naravade (2013). "The Case of Maharashtra's Disappearing Water". *Economic and Political Weekly*, May 4.

Das, K. (2006). *Drinking Water and Sanitation in Rural Maharashtra: A Review of Policy Initiatives*. Ahmedabad: Gujarat Institute of Development Research.

Deshpande, Shashank (2009). *Water Account: A Master Key of Village Water Management*. Government of Maharashtra.

Drinking Water and Sanitation Activity Map, National Rural Drinking Water Programme, MoDW&S, GoI.

Gawade, Vijay (2012). *Maharashtra Sector Status Report—Water Supply, Water and Sanitation Program*. World Bank and Government of Maharashtra, April.

Government of India (2010). *Movement towards Ensuring People's Drinking Water Security in Rural India. Framework for Implementation*. Rajiv Gandhi National Drinking Water Mission, Department of Drinking Water Supply and Sanitation, Ministry of Rural Development.

———. (2011a). *Census of India*. Delhi: Government of India.

———. (2011b). *Report of the Working Group on Domestic Water and Sanitation*. Planning Commission.

Government of Maharashtra (2013). *Economic Survey 2012-13*. Directorate of Economics and Statistics, Department of Planning.

Groundwater Surveys and Development Agency (Government of Maharashtra) and Central Groundwater Board, Government of India (2010). *Report on Dynamic Ground Water Resources of Maharashtra (2008-09)*. November.

IRAP, UNICEF and WSSD, Government of Maharashtra (2013). *Report of Promoting Sustainable Drinking Water Supply and Sanitation in Rural Maharashtra, Institutional and Policy Regimes*, February (unpublished).

Ministry of Drinking Water and Sanitation (2011). *Strategic Plan-2011, Ensuring Drinking Water Security in Rural India*. Department of Drinking Water & Sanitation–Rural Drinking Water, GoI.

Mumbai Gram Panchayat Act 1958, Revised Section 49.

Report on M-DAWS, UNICEF and WSSD, 2010.

Report on Water and Irrigation Commission, 1999.

Riddhi Foundation and UNICEF (2013). *A Report on towards Universal Access to Safe Drinking Water in Rural Maharashtra, Role of Improved Monitoring at Different Levels*, June (unpublished).

Rural Development Department, Government of Maharashtra (1999). *White Paper on Drinking Water Program*, July. p.4.

UNICEF and DWSM Latur (2012). *Bottleneck Report of NBA and NRDWP of Latur District*. November (unpublished).

UNICEF and DWSM Chandrapur (2012). *Bottleneck Report of NBA and NRDWP of Chandrapur District*. November (unpublished).

Water Resources Department, Government of Maharashtra (2012). *White Paper on Irrigation in Maharashtra and Future Policy*, Vol. I, November.

World Bank (2012). *Draft Report on Maharashtra Sector Status Report: Water Supply*. Water and Sanitation Program, Government of Maharashtra and the World Bank.

Annexure Table A-16.1

Key Results from Latur and Chandrapur Districts

No.	Indicators	Latur		Chandrapur	
		Baseline	Current	Baseline	Current
		Units	Units	Units	Units
	Preparation of VWSS Action Plans				
1	Preparation of action plan at village level	-	50	-	60
2	Submission of Action Plan copies to GP, *panchayat samiti* and ZP	-	50	-	60
3	Implementation of action plans initiated	-	50	-	60
	Environment Sanitation				
	Human Excreta Management				
1	Number of families	9676	9977	13686	14097
2	Number of toilets	4063	6153	5592	9341
3	Coverage (in percentage)	42	62	41	66
	3.1 < 20 coverage (number of villages)	12	5	10	2
	3.2 20-40 (number of villages)	13	7	18	6
	3.3 40-60 (number of villages)	7	18	14	16
	3.4 60-80 (number of villages)	1	6	5	17
	3.5 Above 80 (number of villages)	12	11	7	18
4	Number of NGP	7	12	5	7
5	No. of GP selected under Eco Village Project	0	26	0	24
	Liquid Waste Management				
1	Number of individual soak pits constructed	565	599	1090	1286
2	Construction of new gutters (no. of villages)	40	0	40	10

contd...

...*contd*...

	Solid Waste Management				
1	Number of vermi compost	28	49	3	4
2	Number of NADEP	1	1	6	6
3	Number of compost pits	2154	2154	2812	2812
	O&M and Water Quality Plan				
	Water Supply Scheme				
1	Repairing of borewell	0	4	0	10
2	Repairing platform of hand pump	0	1	0	24
3	Repairing of hand pump (mechanical)	0	21	0	34
	Institutional Management				
1	Appointment of Water person	48	50	58	59
2	Training for water person on O&M	40	50	41	58
3	Training of VWSC	0	50	0	60
	Financial Management				
1	Water tax recovery (in percentage)	36.46	64.31	62.72	70.68
	1.1 <40 (number of villages)	23	23	8	0
	1.2 40 – 60 (number of villages)	17	9	19	12
	1.3 60 – 90 (number of villages)	5	32	20	36
	1.4 90 – 99 (number of villages)	0	1	3	5
	1.5 100 (number of villages)	0	1	5	2
	Water Quality Monitoring				
1	Maintain consistency in chlorination of water	37	50	57	58
2	Proper storage of TCL powder	0	50	0	60
3	Availability of TCL powder	48	50	57	60
4	Maintain TCL register regularly	45	50	59	51
5	Maintain OT register regularly	26	11	33	51
	Water Management Plan				
1	Installation of rain gauge	8	49	2	50
2	Maintaining consistency in rainfall recording and ground-water level monitoring	7	36	2	50

contd...

...contd...

3	Households practising organic farming (number of farmers)	0	16	0	69
4	Repairing of water conservation structures	0	3	0	81
5	Construction of new water conservation structures	0	26	0	84
6	Adoption of micro irrigation practices	0	63	0	69
7	Deepening/desilting of water conservation structures	0	1	0	39

Disclaimer: The views expressed herein are those of the authors and do not necessarily reflect the views of UNICEF or the United Nations or GoM.

Acknowledgement: The authors are deeply indebted to Shri Rupindar Singh, Director, Groundwater Surveys and Development Agency, Pune, for his valuable and timely guidance and constant encouragement. The authors are very much thankful to Shri Suresh Khandale, Additional Director, Groundwater Surveys and Development Agency, for his valuable suggestions and to pursue this activity in the state.

17

A.A. CRONIN and B.S. BISHT

Conclusions and Moving Forward

Introduction

This volume is a unique contribution to the governance discourse in India that has implications beyond the water, sanitation and hygiene (WASH) sector. Acknowledging a problem is not easy but is essential if solutions are to be found. This book acknowledges the issues and highlights that such problems are many, complex and intertwined and are easy to become bogged down in. But problem identification, though important, is not the purpose of this book. This book, as outlined in Chapter 1, aims to find solutions and to strengthen governance in rural India. These solutions are applicable just in the WASH sector but can influence all sectors.

The basic question is if it is possible to achieve stronger decentralised WASH governance in India. The book points to a resounding yes! WASH is a wonderful but still underutilised entry point for broader development. Good governance is about building trust and empowering communities to find their own solutions. Practically, this means for Indian rural communities also, there must be a realisation that governance is a partnership between state and citizens. The government cannot provide everything nor should they be absolved of their responsibilities to the community. The community themselves also realise that sustainable water sources and sustained usage of sanitation facilities is partly their responsibility too, otherwise the original social dilemma continues to persist, i.e., why should I contribute to maintain a common pump or use a toilet when no-one else does. This will require a significant mindset change with new scripts to be considered for rural India but efforts can build on the fact that there are positive examples of communities being supportive to each other and that there is a long history of cooperation in India for the common good, as per the Gandhian philosophy. Hence, it will need conflict with deeply held beliefs. This can be said even in spite of the caste and religious splits in India. However, as Gram Vikas' work in Odisha (Chapter 16) shows WASH can change these skewed caste and gender dynamics for the better—now that is improving governance.

In India, there is a strong need to create new WASH social norms around such principles—it must become unacceptable to defecate in the open or not support the maintenance of a public water source in the village as such an act would be an affront and/or a public health threat to all in the village. Recategorisation and changing such expectations may be a relatively long process and it involves building trust as well as some public pressure. There must be collective deliberation on common values and once all are convinced on the way forward, pledges help to disseminate common knowledge of what the group is going to do and expects everyone to do as well as staking their honour on not returning to old ways. Hence, all must know and all must know that all others know that this is the new way forward. This is essential to overcome pluralistic ignorance whereby a majority may not agree with a social practice but may not realise that others are thinking in the same way (Cronin, 2011). This is stressed here as this is holding back improved governance and increasing the reach and impact of the WASH sector.

Transparency and accountability, hallmarks of good governance, are key to this result. But external inputs are also required, in terms of genuine deliberation and triggering of realisation in the disconnect between practices and core values—we practice open defcation but we also want our kids to grow healthy and develop. These are needed for the community to realise that different payoff values would benefit all of them enormously. Secondly, given the collective nature of social norms, the interventions have to reach the entire group so that collective behaviour change is adopted. This means that regular and accepted current 'scripts' or situations need to be changed—at all levels from citizen to ward member to *sarpanch* to district collector to minister. We will return to this point in the Way Forward Section but first we consolidate the key learning and messages from the chapters in this volume.

Key Messages from the Chapters

Bhaskar (Chapter 2) sets the scene for the book in first examining the decentralisation of governance process in India to date and defined the key concept as to build the capacity of the social system to mobilise available energies in a coherent way. Kaur and Kumar (Chapter 7) postulate that it was Gandhi's advocacy of village democracy that laid the foundation for grass-roots level democracy in modern India and describe how in 1992, a landmark bill was passed as the 73rd Constitutional Amendment Act that ensured the transfer of power from the state to the Panchayati Raj Institutions (PRIs).

Bhaskar (Chapter 2) states that in order to be effective, sustainable and acceptable to the local population, such governance must be transparent, open, participatory, communicative, coherent, efficient and above all, accountable. However, with accountability comes responsibility and this is a key problem in the WASH sector in India today—nobody is clear where the accountability lies. This accountability gap then results in poor or no operation and maintenance (O&M) of schemes, poor quality of water and no sustainability of service. Bhaskar also points out that decentralisation helps service delivery to improve as the authority for change move closer to the consumer and calls for funding releases to be linked to incentivising states to do this. This is a point reiterated by Mommen and More (Chapter 10).

O&M is key to sustained systems and a strong reflection of the overall state of governance. Cronin and Khosla (Chapter 3) touch on this point and commend the increasing emphasis in the WASH sector in India on moving away from only capital cost outlays to increasing the focus on sustainability and O&M. Although this has been a positive development from financial and sustainability points of view, Cronin and Khosla (Chapter 3) make the point that this enhances the need for strengthening the human capacity needs (at local and higher levels). The issue of capacity is further developed by Srivastava (Chapter 14).

Mommen and More (Chapter 10) comment that as different government departments are responsible for infrastructure creation, augmentation, renovations and water quality monitoring while the *panchyat* is responsible for O&M which is not logical as the capital and O&M costs are interconnected and strongly influence each other. Choices made in infrastructure creation affect the expenditures in O&M, while lack of regular O&M inflates the costs of renovations. Hence, it is advisable to calculate the total life cycle costs for all newly created hardware. They state that financing of O&M costs is thus ill informed with anecdotal evidence suggesting that sufficient funds are not allocated for O&M and call for an analysis of where public-private partnerships (PPP) may bring increased efficiencies to the sector.

Cronin and Khosla (Chapter 3) comment that the lack of governance is usually manifested by practices ranging from procedural irregularities to illegal and criminal activities, i.e., corruption of various forms. Corruption is the antithesis of good governance and hinders sustainable and equitable development and has been described as being widespread in the WASH sector in India (Davis, 2004). Cronin and Khosla (Chapter 3) call for a tri-prong approach to combat corruption—an integrated approach with three types of audit is needed. Financial auditing is in practice, but when it is done alone, it cannot comment on value for money or on quality of service delivery. It must be linked with a social audit to reflect on what has been delivered to the community. However, even these combined are not enough—a technical audit is required in parallel with financial audits to show value for money, sustainability and safety of the scheme in the long term. The authors argue that it is the lack of this three prong approach of combining financial, social and independent technical auditing that has resulted in many well-intentioned initiatives at curbing corruption not realising their full potential [e.g., Right to Information (RTI) Act, 2005, community agitation, public interest litigations, etc.] in terms of WASH sustainability. Davis (2004) quotes that among successful interventions to combat corruption in South Asia, two traits were common—changing accountability networks in service provision and changed the attitudes of service providers in a way that increased the moral cost to them of misconduct.

Such lessons need to be taken seriously to strengthen WASH governance in India. Mommen and More (Chapter 10) quote the *World Development Report* (2004) where two routes of accountability in the service delivery of water and sanitation services are presented: the long route and the short route. The long route is indirect, layered, tedious and inherently slow and complex involving politicians/policy makers. In the short route, accountability is established directly between the service provider and clients. The National Rural Water Sector framework has a section on community monitoring and social audits. The section specifically mentions monitoring of consumer satisfaction. The short route is therefore recommended to be strengthened, as it facilitates establishment of a direct linkage between the service providers and consumers.

Lakshmi (Chapter 5) notes, like Cronin and Khosla (Chapter 3), that basic strategies to achieve water governance must include systemic risk identification mechanisms to recognise likely areas of corruption; these can be addressed then by strengthening linkages across all the stakeholders for increased transparency and access to information. This calls for systematic application of social accountability mechanisms already proposed RTI Act and the proposed Citizen's Right to Grievance Redress Bill. Ultimately, this can result in terms of leveraging accountability *via* tools like, social audits, community score card, citizen report cards and periodic feedback mechanisms coupled with third party evaluations for quality control. Finally mechanisms of better service delivery may be introduced within the existing service delivery system by stipulating standards and quality benchmarks at different levels of governance, though this requires improved monitoring and evaluation (M&E) systems.

So what needs to change? First of all, the system must trust and nurture the local government system if it is expected to deliver and with quality. Joshi and Sethi (Chapter 6) comment that though O&M is now the responsibility of the local *panchyats*, funds and functionaries have not been transferred—implying a lack of confidence at the political level in terms of capacity of Panchayat Raj department to successfully maintain these schemes. Indeed Kaur and Kumar (Chapter 7) found that although the *sarpanchs* were aware of the exact situation of their village, they were unable to do anything as they had little autonomy. However, experience shows that wherever elected representatives had been given power, they had performed well. Both papers call for key ways forward—capacity building (quantity and quality) and change management interventions with department officials and systemic communication strategy for sustained behaviour change.

A critical issue is of course a thorough technical understanding of the resource. Laksmi (Chapter 5) highlights that water is the key element of rural India influencing all rural productions in agriculture, industry, energy and daily household usage, and in management of the ecosystems. It thus has to be managed rationally and its multiple usages prioritised. Further, its availability and access has to be distributed fairly and transparently, for good governance sake, regardless of the strata that they belong to precisely because the needs are the same. He introduces the term of 'inclusive water governance.'

Structural challenges underpin the WASH and governance landscape in India. Srivastava (Chapter 14) conjectures that while there is a legal framework, it is not being used in a manner which is conducive to decentralisation. On the contrary, the workload on the *gram panchayats* (GPs) is affected due to pressure by the two key stakeholders of the GP—the citizens and the state. They have little powers to negotiate and enter into contracts with different government and non-government institutions to improve service delivery and governance. A strong institution needs to have well-articulated and commonly understood vision and mission, organisation structure and skills, laid down processes and systems, resources and finally incentives to perform, but this is lacking currently. A key current limitation is that accountabilities are unclear. The author worked with two GPs on developing their vision and mission, understanding the current processes. The key input to the vision and mission were pictures drawn by participants of their dream village. They then articulated a vision statement to achieve this. Participatory rural appraisal (PRA) exercises helped developed an action plan and a proposed structure to implement this as well as elaborating various roles and responsibilities. This handholding and mentoring of the GPs in this respect improved service delivery and revenue collection showing above all progress can be made with correct process and capacity.

It is clear, though despite the many challenges, there are many success stories in India around improved governance in the WASH sector in India. Nair and Radhakrishnan (Chapter 4) describes two examples of successful programme implementation backed by governance from Tamil Nadu—the first in terms of urban water management and rain water harvesting in Chennai where success was achieved adopting supply and demand responsive measures backed up by intensive information, education and communication (IEC) campaigns. The second was in terms of convincing householders to accept using waterless toilets in the small municipality of Musri Town Panchayat. Active citizens' participation was the key to both successes, a theme throughout the book.

This is echoed by Lall (Chapter 8) who describes the Uttarakhand Rural Water Supply and Sanitation (URWSS) project in Uttarakhand. This has demonstrated the success of a demand-driven community participatory approach by forming and empowering the communities to plan, construct operate and maintained their own rural water supply schemes. Here the state transferred the control of 'funds functions and functionaries' down to the lowest rung of the local governance infrastructure—GPs—transforming the way rural water supply infrastructure is created and managed in the state. The key has been water supply managers and policy planners, moving beyond water as a technical issue, but one that has social, community, gender, financial, institutional, political, management and environmental aspects as well. Lall sums it up that this success has come once the role of the state government has changed from that of a 'service provider' to that of a 'facilitator.' Likewise, Kumar and Mittal (Chapter 12) call for greater PRI empowerment for sanitation acceleration in Uttar Pradesh (UP).

Goyal and Singh (Chapter 9) state that sustainable solution lies in the management of both the supply and the demand side and identify four key areas for increasing sustainability, including source sustainability (ensuring availability of potable drinking water in adequate quantity throughout the year), system sustainability (optimising the cost of production of water, devising proper protocol for O&M, capacity building of PRIs and awareness generation), financial sustainability (proper utilisation of funds), and finally social and environmental sustainability (proper project management and involvement of all key stakeholders).

These are government stories but also success has been achieved by NGOs also. The experience of Gram Vikas in Orissa (Chapter 15) shows that, the practice of exclusion—of Dalits, *adivasis*, lower castes, widows and women—harms local governance and development. But open defecation-free villages require 100 per cent participation of the community and so this necessitates a complete overhaul of the outlook of the community towards exclusion. Hence, Gram Vikas builds into their programmes non-negotiable core values which include inclusion of ALL families regardless of caste, social or economic status. Gram Vikas assists each village to set up a village general body, represented by every family in the village and members comprising of each head of household, male and female. The second core value is that social and gender equity is a pre-requisite in all self-governance activities. Thirdly, there must be built-in mechanisms for financial, social, environmental and institutional sustainability, and finally all the community must contribute in cash or in kind (construction materials) and labour to develop ownership of the assets created. Again process and product are stressed, not one or the other, and that proper process can lead to long-term impacts on gender, equity and empowerment, a key point reiterated in the literature (see also Lala *et al.*, in press; Biswas, 2012) but sadly the WASH sector in India still lacks a systematic approach to process—a fact compounded by poor governance and corruption.

Mishra *et al.* (Chapter 11) describe successes in terms of community mobilisation around water quality issues leading to local level action. Similarly Hemani *et al.* (Chapter 13) describe how participatory approaches around water safety and security led to the development of water security plans at the village level, with communities at the core of the planning process. Village Action Plan (VAP) preparation included various participatory techniques such as focus group discussions (FGDs), household and institution-level surveys using pre-tested questionnaires, transect walks, survey and mapping of status of household water sources including sanitary surveillance, water quality testing and the status of O&M of available water sources. The VAP development process also took into account the current water availability and consumption pattern, existing schemes, identification of possible additional sources to augment availability of potable water, community perceptions regarding their water sources and quality, water demand, and measures to ensure sustainability. The field challenges included lack of awareness among communities on the importance of water safety and their lack of trust on government functionaries, reflecting on poor governance and supported by defunct water supply schemes, poor O&M as well as lack of ownership. The VAP process showed a clear need for the comprehensive capacity building of *panchayat* representatives and that decentralised governance is a core element for the successful execution of the Water Security Pilot Programme (WSPP).

Kabir *et al.* (Chapter 16) describe similar processes. Even in Maharashtra with comparatively more experience and developed systems at local level, challenges exist. Sustainable village water safety and security (VWSS) processes were started with technical, financial and institutional corrective action identified. Transparent and participatory approaches have helped increase water tax collection by 30 per cent points on an average from the base figure of 40 per cent, sanitation coverage increased and 50 per cent of the villages became open defecation free, all the GPs adopted different kinds of source-strengthening measures, improvement in the storage of bleaching powder, regular chlorination, maintaining registers and all the GPs identified observations wells and installed rain gauges. However, mobilising government officials was a major issue with the authors reflecting also on the need for participatory processes also with officials to mobilise the necessary resources.

Embedding systemic approaches like VWSS in the mainstream government programmes, rather than stand-alone capacity building interventions, was found to enhance the chances of effective governance at village level.

Moving Forward

India's water future is in peril if current management practices and trends persist; the magnitude of the challenges is daunting and a lack of a sharp vision to address them has hindered progress (Cronin *et al.,* in press). This review makes three recommendations that tie in well with the experiences of this volume:

- Government can make increased efforts harness community participation, especially for sanitation and safe, sustainable water supply.
- A radical shift is needed by society and local decision makers to recognise the right to equitable services, especially to ensure improved water and sanitation services for women and girls.
- Institutional reshaping and strengthening is urgently required with a much stronger institutional and policy regime for the new challenges that the India's water sector is facing. This must result in a management and enforcement regime that is evidence-driven.

This, along with renewed focus on capacity building at local level, is in essence, also the key findings from this volume. Shah (2012) echoes these calls for participatory approaches in the WASH sector for strengthened impact. Nayar and James (2010) concluded that such approaches actually benefit/increase user fee collection rates.

Evidence is essential for the key components of effective governance—accountability and transparency. India needs to rapidly and systematically strengthen the WASH M&E and evidence base and move into new areas such as formatives to achieve improved understanding of people behaviours and aspirations prior to implementation. If this evidence is to inform more equitable service provision, it must be disaggregated to capture key social groupings and be gender sensitive. The complexity of water governance outlined in this volume is not unique to India (e.g., Iza and Stein, 2009; Neef, 2009; Moore, 2013) but solutions tailored to the Indian situation are needed, of course building on global learning. The policy opportunities that the 73rd Amendment must now be followed up with are: (i) recommendations of this volume in terms of institutional strengthening and role clarification, (ii) capacity building, (iii) strengthening and diversifying auditing to address corruption, and finally (iv) strengthening of the participatory approaches and interlinking these at all levels.

Ultimately good governance is key to the realisation of the human right to water and sanitation and its indictors in terms of availability, quality, acceptability, accessibility and affordability that must be based within an accountability framework (Rural Water Supply Network, 2013). De Albuquerque (2012) states that the long-term realisation of the rights to water and sanitation demands a broader culture of accountability, which can only flourish in a climate of good governance, strong democratic institutions and transparency. However, De Albuquerque also warns that decentralisation in this respect must be done carefully in terms of real community engagement and the avoidance of increasing power to local elite groups. Neither does it mean higher levels of power may escape the accountability of capacity building and system strengthening.

This volume ultimately aims to move forward the discourse on WASH and decentralisation and good governance in India. The challenges have been outlined but solutions are also available—provided the administrative, political and local community bodies are all aligned to the same goals of improved service delivery. India has many excellent examples and best practices which now need acceleration and scaling-up and move away from islands of excellence to scale provision.

References

Biswas, A. (2012). "A Framework for Rural Drinking Water Quality Management: Collating Experiences from the Voluntary Sector", *Learning Document Issue* No. 3. Bengaluru: Arghyam.

Cronin, A.A. (2011). "Changing Social Norms around the Sustainability of WASH Services in India", University of Pennsylvania Social Norms Dissertation Series. pp.16 (unpublished).

Cronin, A.A., A. Prakash, S. Priya and S. Coates (in press). *Water in India: Situation and Prospects—The Results of a Recent Systematic Sector Review and Consultation*. Water Policy.

Davis, J. (2004). "Corruption in Public Service Delivery: Experience from South Asia's Water and Sanitation Sector", *World Development* 32(1): 53–71.

De Albuquerque, C. (2012). "On the Right Track: Good Practices in Realising the Rights to Water and Sanitation". Portugal: Regulatory Authority for Water and Waste Services (ERSAR). Available on *http://www.ohchr.org/srwaterandsanitation* [accessed May 6, 2013].

Iza, A. and R. Stein (2009). *RULE–Reforming Water Governance*. Gland, Switzerland: IUCN. pp.130.

Lala, S., A.A. Cronin, M. Basu and Jyotsna (in press). "Collating Field Experiences to Inform Structured Approaches for Gender and Equity in WASH in India".

Moore, M.L. (2013). "Perspectives of Complexity in Water Governance: Local Experiences of Global Trends". *Water Alternatives* 6(3): 487-505.

Nayar, V., A.J. James (2010). "Policy Insights on User Charges from a Rural Water Supply Project: A Counter-intuitive View from South India", *International Journal of Water Resources Development*, 26(3): 403-21.

Neef, A. (2009). "Transforming Rural Water Governance: Towards Deliberative and Polycentric Models?" *Water Alternatives* 2(1): 53-60.

Rural Water Supply Network (2013). "Human Right to Water: What does it mean in Practice?" Rural Water Supply Network *Field Note* No. 2013-2: Equity and Inclusion.

Shah, M. (2013). "Water: Towards a Paradigm Shift in the Twelfth Plan", *Economic & Political Weekly* xlvii(3), January 19.

World Bank (2004). "Making the Services Work for Poor People", *World Development Report*. Washington: World Bank.

Disclaimer: The views expressed herein are those of the authors and do not necessarily reflect the views of UNICEF or the United Nations.